Introductory Quantum Physics and Relativity

INTRODUCTORY QUANTUM PHYSICS AND RELATIVITY

JACOB DUNNINGHAM
University of Leeds, UK

VLATKO VEDRAL
University of Oxford, UK
& National University of Singapore, Singapore

Imperial College Press
ICP

Published by

Imperial College Press
57 Shelton Street
Covent Garden
London WC2H 9HE

Distributed by

World Scientific Publishing Co. Pte. Ltd.
5 Toh Tuck Link, Singapore 596224
USA office: 27 Warren Street, Suite 401-402, Hackensack, NJ 07601
UK office: 57 Shelton Street, Covent Garden, London WC2H 9HE

British Library Cataloguing-in-Publication Data
A catalogue record for this book is available from the British Library.

INTRODUCTORY QUANTUM PHYSICS AND RELATIVITY

ISBN-13 978-1-84816-514-4
ISBN-10 1-84816-514-5
ISBN-13 978-1-84816-515-1 (pbk)
ISBN-10 1-84816-515-3 (pbk)

Printed in Singapore.

Preface

This book is based on lecture notes for courses we have taught at the University of Leeds over the past four years. In that sense, it has already been 'battle-tested'. We are very grateful to all the students who took these courses. Their enthusiasm was infectious and made the courses a joy to teach. They also helped by trying out the exercises that appear at the end of each chapter, spotting errors and suggesting changes.

Quantum physics is a fascinating subject and we hope that, in some small way, we are able to convey the wonders and beauty of it to the reader. In turn, we would like to thank the many people who helped spark our interest in the subject, particularly Keith Burnett, Artur Ekert, Peter Knight, Martin Plenio, Anton Zeilinger and the late Dan Walls.

The staff of Imperial College Press have helped enormously with taking our raw manuscript through to the finished book you see today. We would particularly like to thank Laurent Chaminade, Sarah Haynes and Lizzie Bennett for all their support and helpful comments.

Finally we thank our wives Catherine and Ivona and our children Edward, Mikey, Mia and Leo for so many things, but especially for the joy and pleasure they bring to our lives.

Jacob Dunningham
Vlatko Vedral

Preface

Contents

Chapter 1

Introduction

Modern physics is founded on two very important discoveries of the last century: quantum physics and relativity. Both of these were developed to deal with major failings of Newtonian physics. One was in the domain of small particles, such as atoms, where Newtonian physics gives completely wrong predictions. The other failure was in the domain of large velocities, where again Newtonian physics cannot explain some very basic experimental results. The first part of this book will be dedicated to quantum physics; the latter part will introduce the theory of relativity. Both theories have radically changed our everyday intuitions of what objects ought to be and how they ought to behave. It has taken physicists a long time and a great deal of effort to come to terms with some of their implications. A few of these are still viewed as controversial and difficult to grasp even though there is overwhelming experimental support in their favour.

What this book aims to convey are the most recent and accurate descriptions of nature. In some sense they are very different to anything we have seen before. In another very important way, they are not. Quantum physics and relativity are a natural outgrowth of humanity's two a half millennia old quest to understand the universe in a particular rational way. The principles that were laid down by the Ancient Greek philosopher Socrates, the 'grandfather' of Western philosophy, were[1]:

(i) Knowledge can be pursued and is worth pursuing;

(ii) The search for knowledge is a cooperative enterprise;

(iii) A question is a form of education that draws out what is in a person rather than imposing on him a view from outside;

(iv) Knowledge must be pursued with a ruthless intellectual honesty.

You will hopefully begin to appreciate these principles more and more as you progress with your studies. Physics, perhaps more than any other human

[1]These principles have been taken from the book *Socrates* by W. K. C. Guthrie (Cambridge, 1971) [1].

activity, embodies this age-old quest for knowledge. This was, in fact, first fully appreciated by the philosopher Karl Popper. We discuss his ideas a bit here, because they provide a perfect setting for introducing two such revolutionary theories as quantum physics and relativity.

Popper realised that physics (and science in general) makes progress via the method of conjectures and refutations. What this means is that scientists make a conjecture about how something works and then test it in practice to see if their hypothesis is confirmed. If not, the conjecture is discarded and another one tested. It is clear that this will have to be a cooperative enterprise as well as an open minded and honest quest. No matter how fond scientists become of their pet theories, they still have to abandon them when a contradiction is found with experiments. In this way there is a curious philosophy behind any science. No scientific knowledge can ever be proven right. Any theory, if not falsified by an experiment today, just lives to die another day. Another philospher, David Hume, captured this with a now-famous statement (which we are slightly paraphrasing here): "no number of sightings of white swans can ever prove the hypothesis that all swans are white, but observing a single black swan can and does completely invalidate it". And so it is with Newtonian classical mechanics. In this case, there were two black swans of classical physics. One was the theory of relativity and the other quantum physics. By the end of this book (and this is really our main motivation) you will appreciate how quantum physics and relativity together form an important part of our cultural heritage.

We start with quantum theory. Objects in everyday world are considered to be positioned somewhere, to exist at some instant of time and to be traveling at a certain velocity. Newtonian physics stipulates that as soon as we know the position and velocity of an object of a given mass, as well as all the forces that act upon it, then Newton's equations of motion fully determine all the future behaviour of that object. For example, if you know the position and velocity of the Sun, the Earth and the Moon at the present time, you can determine exactly when the next eclipse of the Sun by the Moon will take place on Earth. It is truly remarkable that so little seems to be needed in order to deduce so much. But the compression of all facts into a few simple laws of nature is precisely the point of physics.

Everyday objects, either in addition to or because of the laws of physics, also obey the laws of Boolean logic[2]. It is not clear if logic is something different to physics (probably not!), but let us talk about it as such for the moment. The key law of Boolean logic is the one of excluded middle. An object, such as a chair, is either positioned 'here' or 'not here', but no other possibility can be allowed. If something exists, but it is not here, then it must be elsewhere. And things either exist or do not, no third option is logically possible according to Boole.

[2] George Boole was a 19th century English primary school teacher who was the first person to phrase the laws of thinking in purely mathematical terms – something that baffled and eluded people ever since the ancient Greeks thought about it some 24 centuries earlier. These laws have become the foundations of modern computers and all other forms of information processing.

Another important law of Boolean logic states that if we think of a proposition ("That chair is positioned in this room") and then we think of another proposition ("The chair has zero velocity"), then the order of making these propositions can be reversed and we will still end up with the same overall proposition ("That chair is positioned in this room and it has zero velocity" is simply the same as "That chair has zero velocity and it is positioned in this room"). This law is called the law of commutativity of propositions. You can easily come up with your own propositions and confirm that you can commute them without any change to their meaning.

This all seems fairly reasonable and straightforward. Yet, astonishingly, the world of microscopic objects does not obey Boolean logic and we know this experimentally with a huge degree of certainty. Objects, such as electrons, atoms or photons (particles of light) can, in some sense, exist in many positions (or have many velocities) all at the same time. Furthermore, establishing that "an electron is here now and then that it has some velocity subsequently" is not the same at all as reversing these statements and saying that "an electron has some velocity now and is then located here". In quantum logic, both the law of the excluded middle and the commutativity of propositions are simply violated. How can this possibly be, when these two laws of logic seem extremely fundamental, so fundamental to be beyond any reasonable doubt? We will now present an experiment that illustrates exactly how quantum object behave differently from classical objects[3].

The proof for the existence of particles of light (called photons) has built up over the years since Planck made his 'quantum hypothesis', which we will talk about in detail shortly. Now, however, we want to present a simple experiment to illustrate the basic properties of the quantum behaviour of light[4]. This is meant to motivate the rest of the subject without going into too much detail at this stage.

The apparatus in Fig. 1.1 is called a Mach–Zehnder interferometer. It consists of two mirrors and two beam splitters – these are half-silvered mirrors, which pass light with probability 1/2 and reflect it with the same probability. Let us now calculate what happens in this set up to a single photon that enters it. For this we need to know the action of a beam splitter. It is given by the simple rule[5]

$$\psi_a \rightarrow \psi_b + \psi_c \tag{1.1}$$

which means that the state a goes into an equal superposition of states b and c. This is just a formal way of saying that a particle that enters at path a, exits as

[3]You may be wondering why we keep talking about logic. You might object that logic is a part of mathematics and not physics. It turns out that this is not quite true. Which logic is appropriate in reality is the same question as what the geometry of the universe is. Both are physical questions. Logic and geometry are branches of physics as you will begin to see in this book. One day, it might turn out, that the whole of mathematics is just a branch of physics!

[4]...and also, as it happens, matter.

[5]Note that this state is not normalised. We need a pre-factor of $1/\sqrt{2}$, but since the normalisation is the same for both states a and b we will omit it throughout. We will learn about normalisation a bit later.

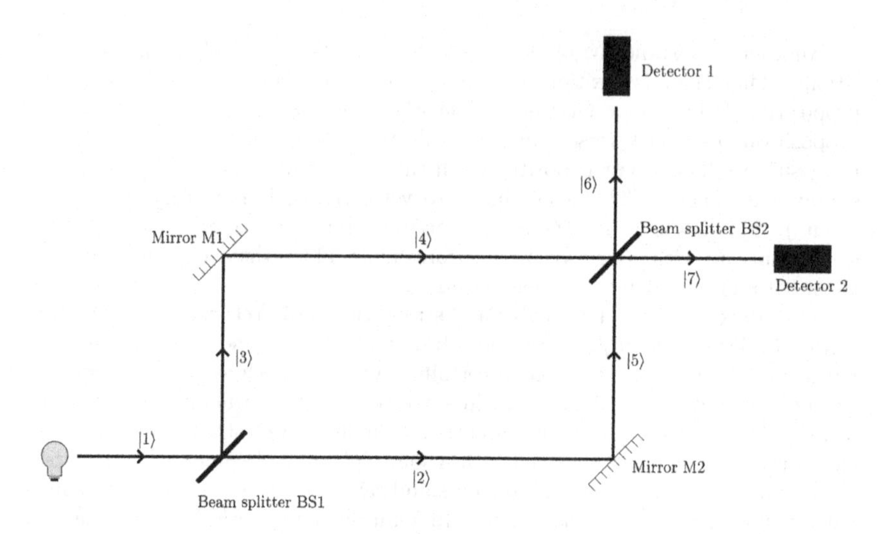

Fig. 1.1. Mach–Zehnder interferometer. This is one of the most frequently used interferometers in the spectral study of light. In this book we will use it mainly to illustrate the 'unusual' behaviour of light in quantum mechanics.

an equal superposition on paths b and c. With this in mind, the Mach–Zehnder interferometer shown in Fig. 1.1 works as follows:

$$\psi_1 \xrightarrow{BS1} \psi_2 + \psi_3 \xrightarrow{M1,M2} \psi_5 + \psi_4 \tag{1.2}$$

$$\xrightarrow{BS2} \psi_7 - \psi_6 + (\psi_6 + \psi_7) \tag{1.3}$$

$$= 2\psi_7. \tag{1.4}$$

Therefore, if everything is arranged properly, and if both of the arms of the interferometer have the same length, then the photon will come out and be detected by detector 2 only[6]. This is called interference and is a well known property of waves – it's just that in quantum physics every photon behaves in this way.

What would happen if we detected light after the first beam splitter and wanted to know which route it took? Then, half of the photons would be detected in arm 2 and the rest of them would be detected in arm 3. So, it seems that photons randomly choose to move one way or the other at a beam splitter. What's more we never detect half the photon in one arm and the other half in the other arm — they are particles and come in chunks. Thus it seems that this is the same as tossing a coin and registering heads or tails. Well, not quite. In fact, not at all as we shall now see. Suppose that at the first beam splitter the

[6]Because we did not normalise the initial state and the states throughout the interferometer, there is an extra factor of 2 in the final result which should be ignored.

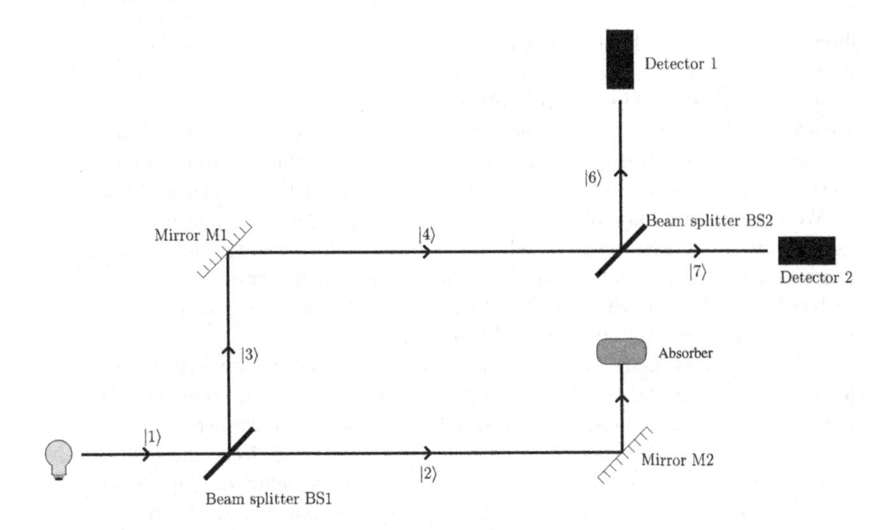

Fig. 1.2. This is the set up involving a Mach–Zehnder interferometer, which shows how strange quantum mechanics is and exemplifies the weird behaviour of quantum objects. The presence or absence of the absorber can be determined without interacting with it. This leads to the notion of the interaction-free measurement – a hot topic in current quantum mechanics research.

photon goes either upwards or to the right, but it definitely goes either up or right (as our experience seems to suggest). Then, at the second beam splitter the photon would again face the same choice, i.e. it would definitely move either up or right. So, according to this reasoning we should expect detectors 1 and 2 to click with an equal frequency. But this is not what we saw. In reality, only detector 2 clicks. The amplitude and, therefore, the probability for detector 1 to click is zero. This means that the operation of a beam splitter and the behaviour of the photon is not just like coin tossing. The state after the beam splitter is more than just a statistical (random) mixture of the two probabilities. It is, of course, a *superposition*, and the photon takes both of the possible routes[7] (in spite of being a particle). This is the true meaning behind writing its state as a mathematical sum of two vectors, say $|1\rangle + |2\rangle$. This is why we use vectors to express states of physical systems.

But we won't be using this more sophisticated description of physical systems until later in the book. There are several good reasons for delaying it. Firstly, the mathematics used for the full quantum theory is quite advanced. Secondly, there are many important features of quantum systems that can be correctly described using less sophisticated theory. Thirdly, starting with simple things and going towards more complicated stuff has a great pedagogical value. It

[7]So photons (and all other physical systems as it turns out) behave according to Yogi Berra's saying: "When you come to a fork in the road, take it!".

shows us how our understanding improves and teaches us never to be dogmatic about our understanding since it is very likely that it will be superseded by some better theory. This is probably the most important part of our scientific culture. And finally, if we started with the complicated theory we would miss out on all the beautiful progress that took place at the beginning of the last century, and it was precisely this progress that made it the 'Century of Physics'.

We close with a very bizarre consequence of quantum mechanics, called the interaction–free measurement[8], which has been performed experimentally using lasers and beam splitters[9]. Suppose that in the Mach–Zehnder interferometer we block one of the paths after the first beam splitter, say path 5, by inserting in there an absorbing material as shown in Fig. 1.2.

What happens then? Well if the photon is absorbed, then neither of the two detectors will eventually 'click' – that's fine. However, if the photon takes the other path, then at the second beam splitter it has an equal chance to be reflected and transmitted so that the two detectors click with equal frequencies. In other words, the interference has been destroyed by the presence of the absorber in path 5. But, here is a very weird conclusion: we can detect the presence of an absorber in path 5 without the photon ever going anywhere near it, hence interaction-free measurement. Thus, if the detector 2 clicks, then the photon has gone to path 6 and that implies that there is an obstacle in path 5 or else only detector 1 would click. This is surely amazing! But that is the basis of the quantum mechanical description of light and all the wonderful phenomena we'll be talking about in this book[10].

[8]This notion was introduced by Elitzur and Vaidman [2]. More details can be found in the paper: Elitzur A.C. Vaidman L. (1993). Quantum-mechanical interaction-free measurements. *Foundations of Physics* **23**, 987–997.

[9]The experiment is reported in the paper: Kwiat P. *et al.* (1995). Interaction-free measurement. *Physical Review Letters* **74**, 4763–4766 [3].

[10]This story can be made all the more dramatic by imagining that instead of the absorber we have a box, which may or may not contain a super-bomb. This bomb is so sensitive that it explodes if a single photon hits it. So to check if the box is hiding the bomb or not we cannot lift the lid as the bomb will then be illuminated and hence will explode destroying the world. Here the Mach–Zehnder set up and the interaction-free measurement come to rescue.

Chapter 2

Old Quantum Theory

Let us begin with the story of how the quantum revolution began. It was Max Planck who first realised that in quantum physics energy has to be quantised. In other words, it comes in small discrete chunks. Before that, physicists believed that energy was a continuous quantity and could have any value, just like most other physical quantities, such as position, momentum and so on. Then Einstein realised a very profound implication of this: light must be composed of fundamental particles called photons. So, light, according to Einstein, is not just a continuous wave, but also has a particle component to it. Finally, de Broglie postulated that matter is not just made up of particles either, but is also wavelike. Electrons are, therefore, not just particles but can and do behave like waves[1].

But, let us not get ahead of ourselves. The whole story starts in 1859, the same year in which Darwin published his *Origin of Species*. Kirchhoff, a famous German experimental physicist, declared that determining the energy spectrum of a black body was the holy grail of physics. He made some preliminary measurements of it, but they were nowhere near precise enough to answer this question. But why did he think that this was so important in the first place, and what do we mean by a black body?

2.1. Black Body Radiation

There are two basic ways in which heat propagates in a given medium: conduction and radiation. Conduction is a relatively simple process. It is governed by a diffusion equation and the rate of change of temperature is proportional to the temperature gradient

$$\frac{dT}{dt} = -\alpha \frac{d^2}{dx^2} T \tag{2.1}$$

where α is a constant that depends only on the properties of the medium, i.e. it does not depend on the temperature. Once the temperature is the same

[1] We will explain exactly what a wave is later on.

everywhere, the right hand side of (2.1) vanishes and there is then no conduction, which is why we said that it was basically a relatively simple process[2]. But there is still radiation, which is independent of any temperature gradient. Radiation processes are much more complicated to study precisely because they can and do take place at constant temperature. Even in equilibrium the behaviour of light is not that simple. So much so that the intense study of its properties led to the advent of quantum mechanics.

Physicists like simple models and extreme situations. Limits of various types abound in physics and as far as radiation is concerned there are two useful situations. One is when we have a body that reflects all radiation that falls on it. This is called a white body. Realistic bodies will of course only strongly reflect in some range of radiation and will absorb for other wavelengths. A mirror is a good example of a white body for visible radiation. The other extreme is a body that absorbs all the radiation that falls on it – a black body. Surprisingly there are a lot of examples of good approximations to black bodies – the Sun and the Earth to name a couple. Suppose that we look at the radiation that leaves the black body. What kind of properties would it have? What would we see if we were inside a black body?

Imagine that you are sitting in your room at night, reading by the night lamp. Your room is not in thermal equilibrium. Why? First of all there are lights on and they emit energy keeping you out of equilibrium. Switch the lights off. Even then, there are objects in your room generating heat and emitting radiation that is not necessarily visible to you. Yourself, for example. You are not in a thermal equilibrium, you generate heat, you are alive. But, suppose that you get rid of all sources and sinks of heat. What properties would the remaining radiation have that is in thermal equilibrium with the matter inside your room?

The remaining radiation would be extremely uniform (the same at every point), isotropic (the same in all directions) and highly mixed, i.e. it would contain many frequencies. Kirchhoff's aim was to describe the spectrum of such a body by finding the energy density as a function of frequency or wavelength.

Towards the end of 19th century it became clear that classical physics could not explain the experimental results of black body radiation. The simple reason was this: classical physics predicts that every atom in the black body should emit radiation at all possible frequencies and that, at temperature T, the energy of each frequency is equal to kT where k is Boltzmann's constant[3]. This energy was classically predicted to be independent of the frequency. Since there are infinitely many possible frequencies, classical physics predicts that the total amount of radiation emitted by a black body is infinite. This prediction is clearly wrong. It is known as the ultraviolet catastrophe since it was a catastrophe for classical physics. The word ultraviolet is there because the density of radiation is quadratically higher at higher frequencies so that the classical prediction becomes worse and worse at higher frequencies, i.e. towards the ultraviolet end

[2]This is oversimplifying this a bit, but nothing too drastic.

[3]Boltzmann's constant can be thought of as the conversion unit between the microscopic energy of particles and their (macroscopic) temperature. It has the value $k \approx 1.38 \times 10^{-23}$ Joules per Kelvin.

of the visible spectrum.

Anyway, by 1900 it was completely clear that classical physics could not explain black body radiation. This was around the time that Planck entered the scene. Now, Planck was already a well established professor who made his name in thermodynamics. He wrote some beautiful papers on it and a few books[4]. However, soon after his work on thermodynamics, he received a book by Gibbs, a famous American physicist, and realised that Gibbs had done all this stuff at least ten years before him. Now if this was not depressing enough, Planck was also trying to prove the second law of thermodynamics when radiation is in equilibrium with matter, but without any success at all[5]. So, disappointed by Gibbs and by his failure to prove the second law[6], Planck decided to attack the problem of explaining the black body spectrum. He spent five years doing this and got nowhere. Then in 1900 he became desperate. He took the best available experimental results of the energy spectrum and extrapolated from it a formula for the energy density. He guessed completely correctly that the energy at a particular frequency is given by

$$E = hf \frac{1}{e^{hf/kT} - 1}. \tag{2.2}$$

Here h is a constant (now known as Planck's constant in his honour) with a value $h \approx 6.63 \times 10^{-34}$Js, f is the frequency of the radiation, k is Boltzmann's constant and T is the temperature of the black body. Note that the overall units correspond to energy since Planck's constant has units of Joules times seconds (Js) and frequency has the units of inverse seconds (1/s), which means that the product has units of Joules, i.e. energy. The rest of the expression is dimensionless and we will understand its exact meaning a bit later.

We can easily show that for high temperatures this goes into the classical limit of kT. At high T, we have $hf/kT \ll 1$ and so the exponential can be approximated as $e^{hf/kT} \approx 1 + hf/kT$. Substituting this into the expression for E above and cancelling the hf factors, we get $E \approx kT$, which proves the classical limit.

The meaning of this quantity is the average energy emitted by black body at the particular frequency. The word average is important simply because the amount of energy fluctuates. This means that at one time you would measure one amount and another time another, but when you average over many measurements you would get the above formula.

We would now like to calculate the total energy output of the black body. For this we need to know how many frequencies there are around the frequency f. Since every frequency corresponds to a different state of black body radiation, this leads to the notion of the density of states. The density of states (per unit

[4]These are classics...

[5]We now know that Planck was probably being a bit hard on himself. No one can prove the second law of thermodynamics from underlying mechanics. This is still one of the greatest open problems of theoretical physics.

[6]and probably reaching the mid-life crisis...

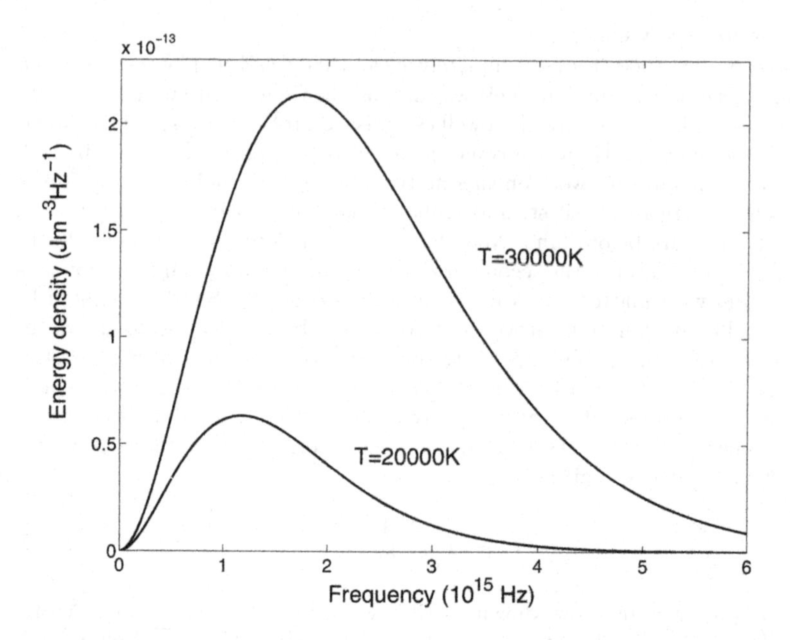

Fig. 2.1. The black body energy density given by Eq. (2.4) for T = 20000K and T = 30000K.

volume) is given by

$$g(f) = \frac{8\pi}{c^3} f^2. \tag{2.3}$$

We are stating this result without deriving it, but the logic is very simple. We can think of different states as lying on a sphere of radius f. The density of states is then proportional to the area of this sphere, which gives us the f^2 factor. Taking into account the fact that there are two polarisations corresponding to each frequency of light gives us the formula above. For now, you don't need to worry about where the other factors come from.

The black body energy density is then given by

$$U = \frac{8\pi h}{c^3} f^3 \frac{1}{e^{hf/kT} - 1}, \tag{2.4}$$

and is shown for two different temperatures in Fig. 2.1. Here $U df$ represents the amount of energy in the interval $f, f + df$ per unit volume. The rationale behind this formula is that it is the product of the density of states with frequency between f and $f + df$ and the average energy per frequency. Before we look at the total energy, which implies an integration over all frequencies, we can already derive one significant conclusion.

It is interesting to consider what wavelength has the highest energy density. This can be calculated by finding the wavelength for which

$$\frac{dU}{d\lambda} = 0. \tag{2.5}$$

First we need to express the energy density in terms of the wavelength. The relationship between wavelength and frequency is as follows:

$$f = \frac{c}{\lambda}. \tag{2.6}$$

This means that

$$df = -\frac{c}{\lambda^2} d\lambda. \tag{2.7}$$

We can ignore the minus sign – all it says is that the frequency and wavelength are inversely proportional, when one increases, the other one decreases and vice versa. Therefore, we get,

$$U = \frac{8\pi hc}{\lambda^5} \frac{1}{e^{hc/kT\lambda} - 1}. \tag{2.8}$$

By looking for the wavelength λ_{max} that maximises this quantity, we derive Wien's law

$$\lambda_{\text{max}} = \frac{0.002897\,\text{Km}}{T}, \tag{2.9}$$

which relates the dominant wavelength of the radiation to the temperature of the black body.

Here is a very beautiful application of Wien's law. The temperature of the surface of the Sun is about 5800 Kelvin. The wavelength of the corresponding radiation is, therefore, roughly $\lambda_{\text{Sun}} = 500$ nanometers (i.e. 500×10^{-9}m). This is in the middle of the visible spectrum of light[7]. The atmosphere scatters the blue component of this light and what is left is yellow, which is why the Sun appears yellow. Classical physics cannot explain the colour of the Sun, but quantum can.

We can also state Wien's law in terms of frequency (though historically he phrased it in terms of wavelength)

$$f_{\text{max}} = 5.789 \times 10^{10}\text{Hz/K} \times \text{T}. \tag{2.10}$$

So Planck found a really beautiful formula, which explained quite a lot of physics known at round about that time. But now he had to somehow try to explain the black body formula. He knew that all the physics in the world at that time (known to us now as classical physics) could not do it. So he decided there was nothing else to do but invent new physics. The 'trick' he employed was to postulate that energy comes in little chunks (quanta) and is not continuous.

[7]The visible light for human beings lies between 400 and 700 nanometers. Yellow is right after green, which occupies the middle of this range at about 530nm. The reason for this is that we evolved in the African savannah where the predominant colour is green and this is what our eyes have adjusted to in the course of evolution.

It is interesting to trace the steps more or less exactly that Planck took towards an explanation of the black body spectrum. The details in the following few paragraphs are really meant for the most interested readers. The argument is fundamentally thermodynamical, which is beyond the scope of this book. Therefore, you will have to just accept some of the things that we state.

So, here is how Planck argued theoretically for his formula. Like we said, he did a bit of reverse engineering. First he guessed the formula from the shape of the curve from the experiment and then he tried to justify it from 'first principles'. Planck knew that thermodynamically speaking, the energy change ΔE for any system is related to its entropy change ΔS in the following way:

$$\Delta E = T\Delta S. \tag{2.11}$$

This is just the conservation of energy if no work is done on or by the system. He also knew that the entropy is equal to the Boltzmann constant k times the logarithm of the number of possible states of the system. This result is the celebrated equation of Boltzmann ($S = k \ln W$), which is a formula engraved on his tombstone in the central cemetery in Vienna[8].

But, now comes Planck's key assumption that revolutionised physics and completely changed the way we view the world. He assumed that the change in energy cannot be smaller than a certain smallest unit, called a quantum, that is

$$\Delta E = hf. \tag{2.12}$$

Now we understand the physical meaning of Planck's constant. It determines the smallest size of the energy packet via the formula above. The entropy is, therefore, also quantised since the number of states available changes in discrete chunks, from say $N+1$ to N quanta. Any smaller change, so Planck stipulated, cannot happen[9]. Putting this together we obtain

$$hf = kT[\ln(N+1) - \ln N], \tag{2.13}$$

where $N = E/hf$. Inverting this expression yields (after a few lines of algebra),

$$E = \frac{hf}{e^{hf/kT} - 1}, \tag{2.14}$$

which matches precisely the form that Planck had guessed to fit the experimental data.

Let us now use this to compute the total intensity of the black body. This implies integration over all frequencies of the quantity

$$\frac{c}{4} g(f) E(f) df. \tag{2.15}$$

[8]This is one of the simplest, most beautiful and most profound equations in science. It links the micro world of W to the macro world of S. This is what allows us physicists a certain degree of arrogance to claim that we can explain chemistry and biology using the fundamental laws of microscopic physics. If the world was about to blow up and we could leave only one message to posterity it would simply have to be this formula.

[9]He had no other reason at the time to suppose this, other than to get the right black body formula.

The physical meaning of this quantity is the energy emitted in the frequency range of f to $f+df$ per unit time and per unit area of the surface of the black body. The factor of 4 comes about by taking into account the fact that the radiation does not always leave orthogonal to the surface and so we need to average over all directions. We can now integrate the quantity (2.15) over all frequencies to obtain:

$$I = \frac{c}{4}\int g(f)E(f)df = \int \left(\frac{2\pi}{c^2}f^2\right) hf \frac{1}{e^{hf/kT}-1}df. \tag{2.16}$$

This integral can easily be evaluated using the identity,

$$\int_0^\infty \frac{u^3}{e^u-1}du = \frac{\pi^4}{15}, \tag{2.17}$$

to obtain the famous Stefan–Boltzmann law

$$I = \sigma T^4, \tag{2.18}$$

where

$$\sigma = \frac{2\pi^5 k^4}{15h^3c^2} \tag{2.19}$$

is the well known Stefan constant. Note that it is not itself a fundamental constant, but is a product of other fundamental constants: the speed of light, c, Boltzmann's constant, k and Planck's constant h. The meaning of I is the energy from the black body that falls on a unit area in a unit of time. The same result can also be obtained by integrating over all wavelengths rather than frequencies:

$$I = \int \frac{2c^2\pi h}{\lambda^5(e^{hc/kT\lambda}-1)}d\lambda. \tag{2.20}$$

You might like to try this as an exercise.

A very interesting property of this formula was discovered by Einstein soon after Planck's revolutionary result. Einstein used Planck's formula to derive the fluctuations of energy and showed that the standard deviation in the black body energy, $(\Delta E)^2 = \langle E^2\rangle - \langle E\rangle^2$, is given by

$$(\Delta E)^2 = hf(N+N^2). \tag{2.21}$$

This can be derived from a thermodynamical relationship that we just quote here: $(\Delta E)^2 = kT^2 dE/dT$. This formula contains both the particle and wave behaviour in one. Let us explain. The particle fluctuations are $\sigma^2 = N$ (the typical $\sigma \sim \sqrt{N}$ deviation for N discretised objects, i.e. particles), while for waves $\sigma^2 = N^2$. Quantum objects, therefore, really behave both like particles as well as waves[10]. Stating it another way, to explain energy fluctuations in a black body it is not enough to assume that black body radiation is made up of particles or waves. We really need the full quantum theory, which somehow merges both.

[10]Or as Sir William Bragg, who did much of his pioneering research at the University of Leeds, famously once said: "God runs electromagnetics by wave theory on Monday, Wednesday, and Friday, and the Devil runs them by quantum theory on Tuesday, Thursday and Saturday". Sunday it seems was anybody's guess.

2.2. The Photoelectric Effect

Unbeknownst to many people, Einstein's first paper quantum theory, although known as a resolution of the photoelectric effect, is mainly concerned with thermodynamical arguments justifying the existence of photons. Einstein proposed an ingenious argument for this. Although very simple, the argument contains thermodynamics that is beyond this book. It is nonetheless worth briefly summarising his argument. Again, you will just have to take some statements on faith.

Einstein reasoned by analogy to 'prove' the existence of photons[11]. He calculated the entropy of a gas of N atoms in a box of volume V. Then he calculated the entropy of a container full of light. The entropy of an atom gas is (we are just stating this result that is given in thermodynamics):

$$S \propto \ln V^N. \tag{2.22}$$

The entropy of a photon gas (also just stated here without derivation) is:

$$S \propto \ln V^{E/hf}. \tag{2.23}$$

So Einstein argued that a gas of radiation is made up of small particles of light – called photons, just as a normal gas is made up of atoms. This, Einstein went on to argue, also resolves some apparent paradoxes that Maxwell's classical theory of light cannot. One of these is the photoelectric effect.

Before Einstein, the photoelectric effect was considered a big puzzle. The effect involves shining light onto a metallic surface and knocking electrons out of it. Electrons require a certain amount of energy to be kicked out. This minimum amount of energy is called the work function. They also have a range of velocities when they leave the metal that can be very precisely measured experimentally. What was found was that:

- Whether or not electrons were knocked out of the metal was largely independent of the intensity of the light;
- It depended instead on the frequency of light, and below some frequency no electrons were emitted;
- The kinetic energy of the emitted electrons was also proportional to the frequency of light.

This was very surprising because the intensity of light determines how much energy light has. And, one would think that the more energy supplied to electrons, the more they should be able to overcome the metallic work function and shoot out. But this was clearly not the case from experimental results.

[11]The inverted commas are there because we never really prove things in physics. We reason and argue and then test. Proving only happens in the certain world of mathematics.

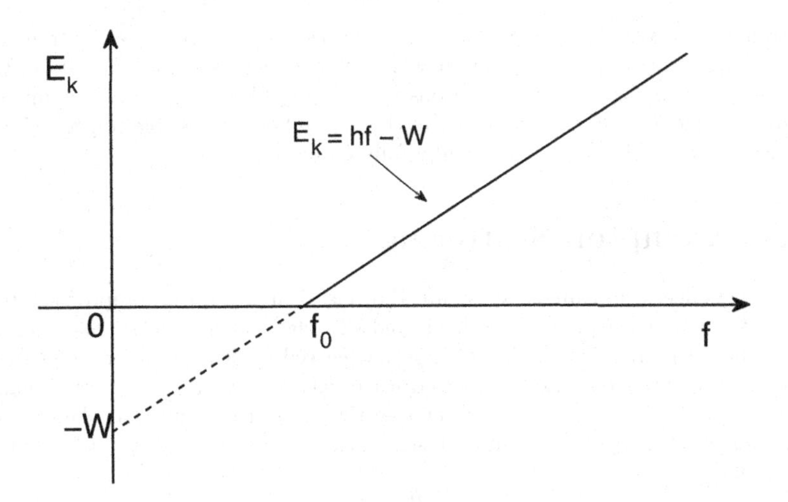

Fig. 2.2. Relationship between the kinetic energy, E_k, of the electrons emitted in the photoelectric effect and the frequency, f of the incident light. There is a threshold frequency below which no electrons are emitted. This is given by $f_0 = W/h$, where W is called the work function and corresponds to the energy needed to knock out an electron.

When explaining the photoelectric effect, Einstein wrote the energy conservation law for photons knocking out electrons from a metal

$$\frac{1}{2}Mv^2 = hf - W, \tag{2.24}$$

which is shown in Fig. 2.2. The left hand side represents the electron's kinetic energy as it leaves the metal, while the right hand side is the difference between photon's energy and the energy needed to knock out the electron. According to this, the photoelectric effect will, therefore, be observed *only* if

$$hf > W \tag{2.25}$$

so that the lowest frequency able to achieve the effect is

$$f_0 = \frac{W}{h}. \tag{2.26}$$

This was Einstein's great success. He used Planck's quantum hypothesis of radiation to explain why the threshold for electron emissions and the velocity of the emitted electrons depend on the frequency rather than the intensity of the light.

As a postscript, we should point out that the photoelectric formula by Einstein is not, in fact, completely correct. If the intensity of light is sufficiently strong, then we can have two or more photons knocking out one electron. So the whole process does, after all, have some dependence on the intensity of the light. These multi–photon excitations are not very likely, but they do happen if there are sufficiently many photons in the beam. So it is possible to get a Nobel prize for something that is not really fully correct[12].

2.3. Compton Scattering

Another effect that cannot be explained by classical physics is Compton scattering. We will first explain what this is and why classical electrodynamics cannot account for it. In 1923, Arthur Compton observed the scattering of x-rays from electrons in a carbon target and found scattered x-rays with a longer wavelength than those incident upon the target (see Fig. 2.3). He found that the shift of the wavelength, $\Delta\lambda$, increased with scattering angle according to the Compton formula

$$\Delta\lambda = \frac{h}{M_e c}(1 - \cos\theta), \tag{2.27}$$

where θ is the angle of deflection of the photon and M_e is the mass of an electron. We will use uppercase M for masses since m will be reserved for another quantity when we come to study the structure of the hydrogen atom. Compton explained and modelled the data by assuming a particle nature for light and applying conservation of energy and conservation of momentum to the collision between the photon and the electron. The scattered photon has lower energy and, therefore, a longer wavelength. We will go through this analysis shortly after showing why a classical treatment breaks down.

In the classical model the electron is treated as an oscillator connected by a 'spring' to a nucleus. Light is now treated as an external force driving the oscillator. This treatment is good at explaining phenomena such as the Rayleigh scattering[13], but fails when it comes to explaining black body radiation and Compton scattering.

To show this we first write Newton's second law of motion for the electron by treating the system as a simple harmonic oscillator

$$\ddot{x} + \omega_0^2 x = \frac{eE}{M_e}\cos\omega t, \tag{2.28}$$

where x is the position of the electron, e is the charge of the electron, E is an electric field, $\omega = 2\pi f$ is the angular form of the frequency of the light and ω_0 is the natural angular frequency of the oscillator. The general solution is

$$x(t) = A\cos(\omega_0 t - \phi) + B\cos\omega t. \tag{2.29}$$

[12]Einstein received his Nobel prize for explaining the photoelectric effect and not for his pioneering work on the theory of relativity. Interestingly, he himself also considered his paper on the photoelectric effect to be his only revolutionary contribution to physics.

[13]which can be used to explain why the sky is blue.

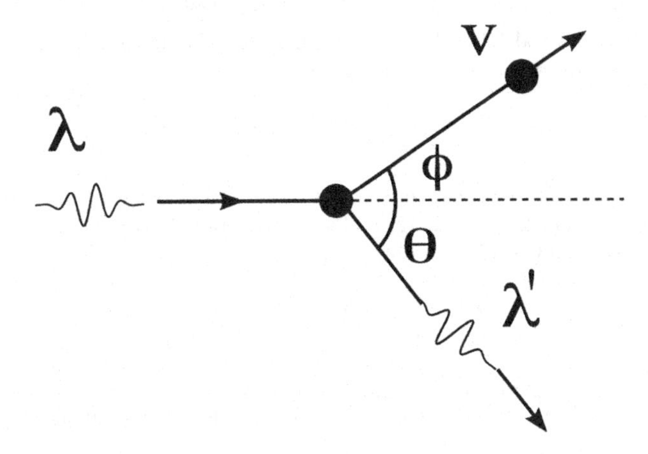

Fig. 2.3. Compton scattering: a photon with wavelength λ is fired at an electron that is initially stationary. After the collision, the photon is scattered at an angle θ and has the new wavelength λ' and the electron is scattered at an angle ϕ with speed v.

Substituting this into the above equation we obtain $B = eE/M_e(\omega_0^2 - \omega^2)$, which gives

$$x = A\cos(\omega_0 t - \phi) + \frac{eE}{M_e(\omega_0^2 - \omega^2)}\cos\omega t. \tag{2.30}$$

The remaining two constants, A and ϕ, can be found from the initial conditions of a particular case. However, we are more interested in some general observations. Firstly, we see from (2.30) that the amplitude is largest when the driving is on resonance, i.e. when $\omega = \omega_0$. Secondly, and more importantly, we see that the atom oscillates at the frequency of the light, i.e. ω. As it oscillates at this frequency, it will also emit light at this frequency. So, there is no way in classical physics that the frequency of light interacting with an electron can change.

Quantum mechanically and in real life, the change in frequency does happen, as Compton was the first to observe. He explained it using the fact that light is made up of photons. Suppose we have a head-on collision. Conservation of momentum implies that

$$\frac{h}{\lambda} = M_e v - \frac{h}{\lambda'} \tag{2.31}$$

where λ and λ' are the initial and final wavelengths of the photon and $M_e v$ is electron's momentum after the collision. We assume that electron is initially stationary and the minus sign is there because they go in the opposite directions after the head-on collision. This can be understood because the momentum of light of wavelength λ is given by the formula

$$p = \frac{h}{\lambda} \tag{2.32}$$

which can be derived from the fact that $E = cp$ for light and $E = hc/\lambda$.

Similarly, conservation of energy gives us,

$$\frac{hc}{\lambda} = \frac{M_e v^2}{2} + \frac{hc}{\lambda'}. \tag{2.33}$$

We can now find λ' by taking the expression for the velocity v from the momentum conservation equation (2.31) and substituting it into the energy conservation equation (2.33). We then obtain

$$\frac{h^2}{2M_e}\left(\frac{1}{\lambda} + \frac{1}{\lambda'}\right)^2 = hc\left(\frac{1}{\lambda} - \frac{1}{\lambda'}\right). \tag{2.34}$$

Tidying up and making the assumption that $\Delta\lambda = \lambda' - \lambda$ is small (so that we can set $(1/\lambda + 1/\lambda')^2 = 2/\lambda^2$ and $\lambda\lambda' = \lambda^2$), we obtain the following formula[14]

$$\lambda' = \lambda + \frac{2h}{M_e c}. \tag{2.35}$$

This coincides with Compton's formula for the head on collision, i.e. when $\theta = \pi$. We see that the final wavelength is, therefore, larger than the initial one. This means that the energy of the final photon is smaller than the initial one. It was simply decreased in the collision with the electron which picked up the difference.

For completeness we will just tell you what you would need to do to derive Compton's formula in general. What changes if the collision is not head on is the momentum conservation equation. The incoming photon momentum need not be in the same direction as the outgoing photon momentum. The angle θ is exactly the angle between the directions of these two momenta (see Fig. 2.3). The momentum conservation equation then looks like

$$\hbar\mathbf{k} = M\mathbf{v} + \hbar\mathbf{k}', \tag{2.36}$$

where the boldface indicates a vector and the magnitude of $\mathbf{k}$ is $k = 2\pi/\lambda$. When $\mathbf{v}$ is squared in this equation and substituted into the energy conservation equation we obtain Compton's full formula providing, again, that we make suitable approximations similar to what we did earlier.

2.4. De Broglie's Hypothesis

From Planck's and Einstein's work it was clear that energy is quantised. So although we think of light as a continuous wave, it is really made up of lots of discrete particles called photons[15]. It was de Broglie, a French prince, who

[14] If you are not used to all sorts of approximations that physicists make, now is the time to start. That is part and parcel of the university degree training for physicists, namely how to cut through all the unnecessary mathematical detail and obtain the desired result.

[15] If you would like to read more about the quantum properties of light, see *Modern Foundations of Quantum Optics* by V. Vedral (ICPress, 2005) [4].

realised that the opposite might also be true. Namely, what we traditionally think of particles, such as electrons, might also behave like waves. He proposed that all matter had wavelike properties. Although simple, de Broglie's idea was so profound that it revolutionised physics.

De Broglie used some very simple formulae to compute a wavelength for a particle of mass M such as an electron; photons on the other hand have zero mass. Here we will retrace his logic. Suppose that the particle moves at the speed v, then its momentum is

$$p = Mv. \tag{2.37}$$

We have seen above that the momentum of a photon is equal to

$$p = \frac{h}{\lambda}. \tag{2.38}$$

If we follow de Broglie and suppose that electrons behave like waves as well, then their wavelength can be obtained by equating the above two expressions

$$\lambda = \frac{h}{Mv}. \tag{2.39}$$

The bigger the mass, the smaller the wavelength. This accounts for the fact that it is very difficult to register any quantum behaviour for a person. We really do not make many waves. The reader may like to try calculating the wavelength of a typical human being assuming 100kg mass and speed of walking of 5m/s. You will find that it is a very small number indeed.

De Broglie's hypothesis was soon confirmed for electrons. In 1927 at Bell Labs, Davisson and Germer fired slow-moving electrons at a crystalline nickel target. The angular dependence of the reflected electron intensity was measured and was determined to have the same diffraction pattern as those predicted by Bragg for x-rays. Bragg's condition for recording maxima in the interference pattern is

$$n\lambda = 2d \sin\theta \tag{2.40}$$

where n is the order of maxima, d is the distance between atoms in the crystal and θ is the angle of diffraction of the electrons.

Before de Broglie's hypothesis, diffraction was a property that was only exhibited by waves. Therefore, the presence of any diffraction effects by matter demonstrated its wavelike nature. When the de Broglie wavelength was inserted into the Bragg condition, the observed diffraction pattern was predicted, thereby experimentally confirming the de Broglie hypothesis for electrons.

This was a pivotal result in the development of quantum mechanics. Just as Compton demonstrated the particle nature of light, the Davisson–Germer experiment showed the wave-nature of matter and completed the theory of wave-particle duality. For physicists this idea was important because it means that not only can any particle exhibit wave characteristics, but that one can use wave equations to describe phenomena in matter if one uses the de Broglie wavelength. We will talk about this more later.

Of course, we need not stop with electrons. According to de Broglie, all matter should behave like a wave and should, therefore, exhibit effects, such as diffraction and interference, that we normally associate with waves. What is the state of the art when it comes to interference of matter? The most advanced interference to date has been carried out in the group of Anton Zeilinger in Vienna with molecules involving 120 carbon atoms. Even these are shown to behave like waves when passing through diffraction grating.

De Broglie's hypothesis can be summarised by the relationships between energy, momentum and wavelength. These are

$$p = \frac{h}{\lambda} = \hbar k \tag{2.41}$$

$$E = \frac{hc}{\lambda} = hf = \hbar\omega \tag{2.42}$$

where $\hbar = h/2\pi$ is just a different form of Planck's constant (called the reduced Planck constant) that you will see a lot more of.

2.5. Bohr's Model of the Atom

The existence of atoms was postulated by the ancient Greeks and was confirmed beyond any reasonable doubt in the scientific community by 1909 with the explanation of Rutherford scattering. However, classical physics could not explain their stability and persistency. Classical mechanics implies that an accelerated charge has to radiate. As it radiates it loses energy and, therefore, has to slow down. An electron orbiting a nucleus would have to lose energy and collapse into the nucleus, but this is never observed.

The original argument for the instability of an atom is due to Lorentz and it requires a very advanced knowledge of classical electrodynamics. Here we follow a beautiful alternative argument proposed by J. J. Thomson[16]. When a charge is stationary the electric field lines originate from the place where the charge is and stretch away to infinity. When the charge is accelerated by Δv in time Δt, then the signal for the change of the field lines will only propagate within a sphere of radius $c\Delta t$ as it propagates at the speed of light[17], c. So the lines outside the sphere will not match the ones within as they have not yet received the signal that the charge has been accelerated in the first place. But this cannot be since the field lines need to be continuous, otherwise it would be non-physical. This means that at the boundary the E field cannot just have a radial component, but must also have a tangential component. This, in turn, means that there is a component of the electric field which is propagating away from the charge and taking away some energy. Finally all the energy is taken away in this manner and so the atom stops oscillating.

[16]Among his many achievements, J.J. Thomson is credited as the discoverer of the electron, for which he won the Nobel prize in 1906. Remarkably seven of his research assistants also went on to win Nobel prizes as did his son George Paget Thomson.

[17]As we will see later in the book, this is the highest speed at which any information can propagate, according to Einstein's relativity.

With the use of spectroscopy in the late 19th century, it was found that the radiation from hydrogen, as well as other atoms, was emitted at specific quantised frequencies. It was the effort to explain this radiation that led to the first successful quantum theory of atomic structure, developed by Niels Bohr[18] in 1913. He developed his theory of the hydrogenic (one-electron) atom from the following four postulates:

- An electron in an atom moves in a circular orbit about the nucleus under the influence of the Coulomb attraction between the electron and the nucleus, obeying the laws of classical mechanics.

- Instead of the infinity of orbits, which would be possible in classical mechanics, it is only possible for an electron to move in an orbit for which its orbital angular momentum L is an integral multiple of Planck's constant $\hbar = h/2\pi$ (this is still an infinity, but countable).

- Despite the fact that it is constantly accelerating, an electron moving in such an allowed orbit does not radiate electromagnetic energy (this is postulated, not derived). Thus, its total energy E remains constant.

- Electromagnetic radiation is emitted if an electron, initially moving in an orbit of total energy E_i, discontinuously changes its motion so that it moves in an orbit of total energy E_f. The frequency of the emitted radiation f is equal to the quantity $E_f - E_i$ divided by Planck's constant h.

Note that Bohr[19] did not really resolve the problem of the stability of atoms, he just postulated that they are stable. This is the way we usually remove difficulties in physics, by simply saying that there are none[20].

Two equations are now sufficient to capture the above postulates. The first is the equation following from the second postulate:

$$L = n\hbar, \tag{2.43}$$

where n is an integer. Classically, the angular momentum is given by

$$L = M_e v r \tag{2.44}$$

[18]Niels Bohr could have become a famous footballer rather than a physicist. He played goalkeeper in the Danish team Akademisk Boldklub, a top Danish team, at the beginning of the 20th century. His brother Harald, a well-known scientist in his own right, played for the Danish national team and won silver at the 1908 London Olympics.

[19]Another interesting story about Bohr is that he was smuggled out of Denmark in a fighter-bomber during the Second World War to work on the Manhattan project. During this escape, he was seated in an improvised seat in the bomb bay of the plane. The flight almost ended in tragedy as Bohr did not wear his breathing equipment and passed out at high altitude. Fortunately, the pilot guessed what had happened and descended to a lower altitude for the rest of the flight. Afterwards, Bohr said that he had slept like a baby for the entire flight.

[20]And this is, of course, in stark contrast to mathematicians.

where r is the radius of electron's orbit. Note that this quantisation condition is the same as the consistency for standing waves

$$n\lambda = 2\pi r_n, \tag{2.45}$$

which postulates that an integral number of wavelengths has to fit into the electronic orbit. Here r_n denotes the radius of the nth quantised orbit.

Now, balancing the centrifugal force and the Coulomb attraction between the electron and the nucleus, we get

$$\frac{M_e v_n^2}{r_n} = \frac{Ze^2}{4\pi\epsilon_0 r_n^2}, \tag{2.46}$$

where ϵ_0 is the permittivity of free space, v_n is the speed of an electron in the nth orbit and Z is the atomic number of the atom, i.e. the number of protons.

From these two we can calculate the radii of the electron orbits, as well as their energy. We obtain for the nth radius

$$r_n = \frac{4\pi\epsilon_0 \hbar^2 n^2}{M_e Z e^2}. \tag{2.47}$$

The radius for which $n = Z = 1$ is known as the Bohr radius (roughly half a nanometer in size). The velocity, v_n, for an electron in the nth orbit is then,

$$v_n = \frac{Ze^2}{4\pi\epsilon_0 \hbar n}. \tag{2.48}$$

As we see, the velocity of electrons goes down with n as we move away from the nucleus.

It is now very straightforward to calculate the total energy in the nth orbit. The total energy is the sum of the kinetic and potential energies

$$E_n = K_n + P_n = \frac{M_e v_n^2}{2} - \frac{Ze^2}{4\pi\epsilon_0 r_n}. \tag{2.49}$$

The potential energy has a negative sign, because the force between the nucleus and the electron is attractive. Another way of thinking of it is that the electron needs some positive kinetic energy to overcome this force.

Using the formulae for the radii and velocities we obtain

$$E_n = -\frac{M_e Z^2 e^4}{(4\pi\epsilon_0)^2 2\hbar^2}\frac{1}{n^2}. \tag{2.50}$$

The minus sign, as before, indicates that the energy is due to an attractive (binding) force. If we take the case of hydrogen in the lowest energy level, i.e. $Z = n = 1$, we obtain

$$E = -2.1 \times 10^{-18}\text{J} = 13.6\text{eV}, \tag{2.51}$$

which is known as Bohr's energy. It is the amount of energy required to ionise a hydrogen atom.

Once the energy difference between the two states is known, and this can be obtained from Bohr's formula, then the wavelength follows from the relationship $E = hc/\lambda$. In this way, the wavelength of the photon emitted when an atom makes a transition from level $n+1$ to n can be shown to be

$$\frac{1}{\lambda} = R\left[\frac{1}{n^2} - \frac{1}{(n+1)^2}\right], \tag{2.52}$$

where $R \approx 1.097 \times 10^7 \mathrm{m}^{-1}$ is the Rydberg constant and is given by

$$R = \frac{M_e e^4}{8\epsilon_0^2 h^3 c}. \tag{2.53}$$

Unfortunately, in the long run the Bohr theory, which is part of what is generally referred to as the old quantum theory, is unsatisfying. Looking at the postulates upon which the theory is based, the first postulate seems reasonable on its own, acknowledging the existence of the atomic nucleus, established by the scattering experiments of Ernest Rutherford[21] in 1911 and assuming classical mechanics. However, the other three postulates introduce quantum-mechanical effects, making the theory an uncomfortable union of classical and quantum-mechanical ideas. The second and third postulates seem particularly ad hoc. The electron travels in a classical orbit, and yet its angular momentum is quantised, contrary to classical mechanics. The electron obeys Coulomb's law of classical electromagnetic theory, and yet it is assumed not to radiate, as it would classically. These postulates may result in good predictions for the hydrogen atom, but they lack a solid fundamental basis.

2.6. Problems with Old Quantum Theory

The results we have presented so far have all been very big departures from classical physics. However, they have not been radical enough to describe all the observed phenomena in the early 1900s. For example, Bohr's model was good at predicting observations for hydrogen, but was very bad for any other elements in the periodic table. More complex systems, such as molecules or liquids and solids were even more difficult to handle quantum mechanically. Secondly, there was a problem of understanding quantum dynamics. How exactly do quantum systems change in time?

The Bohr theory is also fatally incomplete. For example, the Wilson–Sommerfeld quantisation rule, of which the second Bohr postulate is a special case, can only be applied to periodic systems. The old theory has no way of approaching non-periodic quantum-mechanical phenomena like scattering. Next,

[21] Rutherford once remarked that "all science is either physics or stamp collecting". As if by way of revenge, when the Nobel committee awarded him a Nobel prize for his work on radioactive decay it was in chemistry, not physics.

although the Bohr theory does a good job of predicting energy levels, it predicts nothing about transition rates between levels. To correct these faults, one needs to apply a more complete quantum-mechanical treatment of atomic structure. Fortunately, Heisenberg, Schrödinger, Dirac and Born were on hand to develop a general framework to address all these questions and much much more. This sets the scene for the next chapter in which we describe quantum mechanics proper in all its glory.

2.7. Exercises

At the end of each chapter we will provide a number of exercises and problems of varying difficulty. These are meant to aid you in understanding the material presented in the chapter. Practicing problem solving is the key to any understanding and skill in science[22]. Solutions are given at the back of the book.

1. One can estimate when quantum effects become important by comparing the de Broglie wavelength of the system under consideration to its size. In thermal equilibrium the kinetic energy of a system is given by

$$\frac{p^2}{2M} = \frac{3kT}{2}.$$

   Derive the de Broglie wavelength as a function of temperature.

   The lattice spacing between atoms in a typical solid is $d = 0.3$nm. Find the temperature below which the free electrons in a solid are 'quantum mechanical'. What about the atomic nuclei?

   What is, therefore, the appropriate way of treating a solid at room temperature?

2. The de Broglie wavelength tells us also about the spatial extent of the system it corresponds to. This question is about determining the temperature below which we have an effect called Bose–Einstein condensation that you will learn more about in the next chapter. Suppose that the you have N free atoms of mass M and at temperature T. Consider first the one-dimensional case and suppose that the density of particles is uniform everywhere $\rho = N/L$.

   Derive the temperature for condensation of atoms in terms of M and ρ by calculating when the de Broglie wavelength is larger than the typical average distance between atoms.

   Calculate this critical temperature for 10^5 rubidium-87 atoms that are confined inside a one-dimensional trap of size $L = 100\mu$m.

   How would this relationship change in three spatial dimensions and what is the corresponding temperature?

3. This question concerns estimating the temperature on Earth by finding its equilibrium with Sun's radiation. We will address it in the following sequence of steps.

   First, write down the total power emitted by the Sun, as a function of its temperature T_S and radius R_S. Show that the power absorbed by Earth

[22] And not just in science.

(assuming that it is a perfect absorber) is given by

$$P_{Eabs} = \sigma T_S^4 \frac{\pi R_S^2 R_E^2}{D^2}$$

where σ is Stefan's constant, R_E is the Earth's radius and D is the distance between the Earth and the Sun.

Write down the power emitted by the Earth as a function of Earth's temperature T_E and its radius R_E. By equating the absorbed and emitted powers of the Earth derive the formula for Earth's temperature

$$T_E = \sqrt{\frac{R_S}{2D}} T_S,$$

which is a very simple and elegant result.

Is the temperature calculated from above lower or higher than the actual average? What can contribute to this difference and why?

4. This question will address the total human power output and why we do not see humans glow in the dark.

   Calculate the difference between emitted and absorbed powers of a human being whose area is 2m^2. Assume that the temperature of the human skin is about 305K while that of the surroundings is 293K.

   How does this power compare to a 100W light bulb?

   Use Wien's law to calculate the dominant wavelength of human emission. What part of the spectrum does it belong to?

5. Calculate the wavelength of light emitted when the atom makes a transition from the $(n+1)$th to nth level in Bohr's atomic model. Where in the spectrum does this wavelength lie for $n = 1$?

   For which smallest n does this lie in the visible spectrum?

   For the purposes of deriving Bohr's formula we have assumed that the nucleus is stationary, which amounts to assuming that the nuclear mass, M_n, is infinite. How would you argue that to correct this assumption we need to use the following mass

$$\mu = \frac{M_e M_n}{M_e + M_n}$$

   instead of the electron mass M_e. How different is this (as a percentage) from the previous estimate for the $n = 2$ to $n = 1$ transition?

Chapter 3

Quantum Mechanics

We now describe the full theory of quantum mechanics which emerged in the 1920s. There are four rules of quantum mechanics and they fully describe anything that we would like to calculate. These ideas will seem strange at first and it will probably take you a while to feel comfortable with them. Bohr himself said that "those who are not shocked when they first come across quantum mechanics cannot possibly have understood it". It is, however, well worth the effort.

The rules are:

1. States of physical systems are represented by complex functions, called wave functions.

2. Observables of physical systems, things that we can measure, such as the position or momentum of a particle, are represented by Hermitian operators.

3. When we make a measurement of the system we probabilistically obtain the answer according to the modulus square of the wave function. This is the so-called Born measurement postulate.

4. The system, when not measured, evolves according to the Schrödinger equation.

The meaning of each of these statements will become clear as we start to develop quantum mechanics. Suffice it to say that the statements are neither intuitively clear nor physically motivated at this stage[1]. The ultimate justification, as always for a physicist, is that these rules are in extremely good agreement with experimental results and tests. Quantum mechanics is our by far most accurate description of nature. With that in mind, let us proceed. First we will discuss the Schrödinger equation that appears in the fourth rule, which was in fact discovered well before the other postulates of quantum mechanics were phrased.

[1] Some researchers believe that we need a better formulation, but there is no other formulation at present.

3.1. Schrödinger's Equation

If particles are waves then it would seem reasonable to expect to be able to write a wave equation for these quantum objects[2]. So we need to understand how waves behave in the first place.

Waves were already well understood by physicists in the nineteenth century. A wave equation is usually an equation that involves a second derivative in space as well as in time. It reads something like

$$\frac{\partial^2 f(x,t)}{\partial x^2} = \frac{1}{v^2}\frac{\partial^2 f(x,t)}{\partial t^2} \tag{3.1}$$

Here v is the velocity at which the wave propagates. If the above is to describe light, then the velocity is simply the speed of light, if, on the other hand, we are describing sound, then v is the speed of sound and so on.

A solution to this equation is the so-called plane wave:

$$f(x,t) = Ae^{i(kx-\omega t)} \tag{3.2}$$

such that $\omega = vk$. Here, ω is the frequency of the wave oscillations and $k = 2\pi/\lambda$ is its wavenumber. A is known as the amplitude of the wave and for now we will treat it as an arbitrary constant. Its role will be discussed in more detail below. The number i is the imaginary number[3] whose value is $\sqrt{-1}$.

The plane wave $f(x,t)$ is a complex function, i.e. a function involving complex numbers, which are combinations of real and imaginary numbers. In classical physics, however, waves are always real. Let us take the electromagnetic wave as an example. The electric field can be written in the following form (the same goes for the magnetic field):

$$E = E_0 \cos\left(\frac{2\pi}{\lambda}x - \omega t + \theta\right), \tag{3.3}$$

where all the symbols have their usual meaning and θ is the initial phase, which can take any value. In quantum mechanics, as we will now see, waves can have a genuinely complex nature and this is one of the key differences between quantum and classical physics.

[2]The story says that Schrödinger was giving a talk about de Broglie's ideas, when someone from the audience got up and said: "If matter is really made up of waves, then what is the wave equation describing its behaviour?" Schrödinger did not know the answer, but had to work it out over the following weekend. This resulted in his celebrated equation for which he was awarded the Nobel prize in 1933.

[3]The imaginary number was discovered when an Italian medieval mathematician encountered the equation $x^2 + 1 = 0$. He knew that the solution of $x^2 - 1 = 0$ was $x = \pm 1$, but the equation $x^2 + 1 = 0$ did not seem to him to have any real solution. So he introduced an imaginary unit i and the solution could then be written as $x = \pm i$. Imaginary numbers were attributed all sorts of mystical properties after that for a long time. In some parts of Europe they were even banned by the Inquisition and people who promoted them were persecuted. Anyway, as if to show that God had a sense of humour after all, these numbers ended up being indispensable in the most accurate description of the world we have – quantum mechanics.

Let us take an electron, for example, and try to derive a wave equation describing its behaviour. This is actually how Schrödinger arrived at it as well. The electron's kinetic energy is given by

$$E = \frac{p^2}{2M} = \frac{\hbar^2 k^2}{2M}, \tag{3.4}$$

but de Broglie also tells us that the energy should be like that of a wave, which means that we should be able to associate some natural frequency to this electron, i.e.

$$E = \hbar\omega. \tag{3.5}$$

By equating the two we see that

$$\omega = \frac{\hbar k^2}{2M}, \tag{3.6}$$

i.e. now the frequency of the electron is proportional to k^2 (and not to k as in our wave equation). If a wave is to be associated with an electron its plane wave form, i.e. its wave function, would be

$$\Psi = e^{i(kx - \hbar k^2 t/2M)}. \tag{3.7}$$

Now, what differential wave equation in space and time would this be a solution to? We are in the funny position that we know the answer, but now we have to come up with an equation whose solution is the above wave. This was basically the task that Schrödinger had to undertake.

After staring at this for some time, it should become clear that we need a second derivative in space, as before in the wave equation, but, unlike in the wave equation, we need a first derivative in time. This is because we need the equation to hold true regardless of the value of k (or ω) and

$$\frac{\partial^2}{\partial x^2}\Psi = -k^2 e^{i(kx - \hbar k^2 t/2M)} = -k^2\Psi \tag{3.8}$$

while

$$\frac{\partial}{\partial t}\Psi = -i\frac{\hbar k^2}{2M} e^{i(kx - \hbar k^2 t/2M)} = -i\frac{\hbar k^2}{2M}\Psi. \tag{3.9}$$

So if we equate the two derivatives above, the k^2 factors cancel and it is possible to write

$$\frac{-\hbar^2}{2M}\frac{\partial^2}{\partial x^2}\Psi = i\hbar\frac{\partial}{\partial t}\Psi. \tag{3.10}$$

This is the celebrated Schrödinger equation, albeit in a restricted form.

What if the particle is subject to an external force, whose potential is $V(x)$? Then the equation is modified in the following way

$$\left[\frac{-\hbar^2}{2M}\frac{\partial^2}{\partial x^2} + V(x)\right]\Psi = i\hbar\frac{\partial}{\partial t}\Psi \tag{3.11}$$

and this is the full Schrödinger equation in all its glory. We are only stating it in one dimension at the moment, but we will see shortly how it generalises to three-dimensional space. Later in the book we will also revisit this equation and discuss in more detail why it has the form that it does. We will see that it is nothing more than an expression of the conservation of energy written in a formal quantum mechanical way.

So, the most basic equation of motion in quantum mechanics is similar to a wave equation but, unlike the wave equation, space and time enter the Schrödinger equation in an asymetric way. The Schrödinger equation is, therefore, more like a diffusion equation. This suggests that a wavepacket will spread out, or diffuse, with time. We can understand this better by briefly discussing the velocity of the propagation of waves.

There are two important velocities to mention: group and phase. They are defined as follows. The group velocity is

$$v_g = \frac{d\omega}{dk} \tag{3.12}$$

while the phase velocity is

$$v_p = \frac{\omega}{k}. \tag{3.13}$$

The actual physical velocity of wave propagation is always the group velocity. For the free particle we know that

$$\hbar\omega = \frac{\hbar^2 k^2}{2M} = \frac{Mv^2}{2} \tag{3.14}$$

and, therefore, $v_g = \hbar k/M$. The phase velocity, on the other hand, has half this value, $v_p = v_g/2$, for a free particle and is the speed of propagation of the points in phase. In some cases this can even exceed the speed of light.

The following example may help here. Suppose we add two sinusoidal waves whose frequencies and wavenumbers are very close to one another

$$\sin(kx - \omega t) + \sin[(k + dk)x - (\omega + d\omega)t] \approx 2\sin(kx - \omega t)\cos\left(\frac{dkx - d\omega t}{2}\right).$$

We see that the overall wave is just like the original one, i.e. $\sin(kx - \omega t)$, but governed by an envelope, $\cos[(dkx - d\omega t)/2]$, which travels at the speed $d\omega/dk$, which is just the group velocity.

But why is the ψ wave function complex and not real? What does that mean? The imaginary number $i = \sqrt{-1}$ appears in Schrödinger's equation, but it never appears in any classical equations of Newton or Maxwell. This looks very unusual and it certainly puzzled Schrödinger and other people in the community at that time.

3.2. Born's Postulate

It seems that it is very difficult to attribute any reality to the wave function. It is a complex function and complex numbers never really appear in our measurements. We always measure some physical observable, such as the position of a particle, and the outcome is always a real number. Interestingly enough, the meaning of the wave function was not clear to Schrödinger either, even though he had invented it in the first place[4].

The German physicist Max Born resolved this issue in 1925 (for which he was awarded the Nobel prize in 1954), with what has become known as the Born interpretation[5]. He stated that the modulus squared of the wave function is the probability density for particle to be found at a position x[6]. Therefore, the probability to find a particle located between x and $x + dx$ is given by

$$|\Psi(x)|^2 dx \tag{3.15}$$

The probability to find a particle in a finite interval between x_1 and x_2 is consequently

$$p = \int_{x_1}^{x_2} |\Psi(x)|^2 dx. \tag{3.16}$$

A direct consequence of this is that

$$\int_{-\infty}^{+\infty} |\Psi(x)|^2 dx = 1 \tag{3.17}$$

This means that the total probability for a particle to be somewhere must be one. This is often called the normalisation condition – it makes sure all the probabilities add up as they should.

3.3. Time-Independent Schrödinger Equation

We have seen that the full one-dimensional Schrödinger equation for a particle described by the wave function $\Psi(x, t)$ is,

$$-\frac{\hbar^2}{2M}\frac{\partial^2 \Psi}{\partial x^2} + V(x)\Psi = i\hbar\frac{\partial \Psi}{\partial t}. \tag{3.18}$$

Just like Newton's equations, Schrödinger's equation cannot be derived from any more fundamental principle – it is a *law* of quantum physics[7]. The reason we are happy to use it is simply that it works. No experiment has yet shown

[4] This is always the nature of doing something revolutionary – all the consequences are not immediately clear.

[5] Max Born's other claim to fame is as the grandfather of Olivia Newton-John.

[6] What he stated in the text of his paper in 1927 is that the wave function itself gives the probability density, but then he added a footnote in the revised version to say that "on closer inspection it is the square of the wave function that plays this role".

[7] It is sometimes seen as a weakness of physics that it rests on unprovable laws. However, this view denies the scientific process, which is not about proving theories, but rather trying to disprove them. Schrödinger's equation is a scientific theory which has, so far, stood up against a century of rigorous testing, but that does not mean that someday it will not be superseded. The fact that science relies on unprovable laws also suggests that there are things that lie beyond the scope of science.

any deviation from its predictions. Importantly, Schrödinger's equation tells us how the wave function changes with time so, if we know the wave function at a particular time, we can use (3.18) to calculate the wave function at any other time. In order to calculate solutions to the Schrödinger equation, it is useful to first find its time-independent form.

To do this we look for solutions that have the separable form,

$$\Psi(x,t) = \psi(x)\eta(t), \tag{3.19}$$

where $\psi(x)$ is a function of x only and $\eta(t)$ is a function of t only. This method is known as the separation of variables technique (SVT) and is a useful tool for solving differential equations. You will meet it again later in the book when we come to solve the Schrödinger equation describing the electron in a Hydrogen atom. Substituting (3.19) into (3.18) and dividing both sides by Ψ gives,

$$-\frac{\hbar^2}{2M}\frac{1}{\psi(x)}\frac{\partial^2\psi(x)}{\partial x^2} + V(x) = i\hbar\frac{1}{\eta(t)}\frac{\partial\eta(t)}{\partial t}. \tag{3.20}$$

Importantly, the left hand side of the equation depends only on x and the right hand side depends only on t, i.e. the variables have been separated. Since the x and t values are independent, i.e. if we change x or t on one side of the equation it cannot affect the other side, the two sides can only be equal if each is equal to the same constant. We will call this constant E since it turns out to be the total energy of the system.

Solving the equation for $\eta(t)$,[8]

$$i\hbar\frac{d\eta}{dt} = E\eta, \tag{3.21}$$

gives $\eta = \exp(-iEt/\hbar)$. The equation for $\psi(x)$ is,

$$-\frac{\hbar^2}{2M}\frac{\partial^2\psi(x)}{\partial x^2} + V(x)\psi(x) = E\psi(x). \tag{3.22}$$

This is called the time-independent Schrödinger equation. We note that the solution to the full time-dependent Schrödinger equation (3.18) is easily found from the solution, $\psi(x)$, to (3.22), i.e. $\Psi(x,t) = \psi(x)\exp(-iEt/\hbar)$. One important feature of solutions that take this form is,

$$|\Psi(x,t)|^2 = |\psi(x)|^2, \tag{3.23}$$

i.e. the probabilities do not change with time and so are called stationary states. For stationary states, we can use the time-independent form of the Schrödinger equation. This is the equation that we will be interested in for a large part of this book.

[8] $i\hbar(\partial/\partial t)$ is called the energy operator. Operators are a very important part of quantum mechanics and we will discuss them in more detail later in the book.

3.4. Free Particle

We have already considered the simplest solution of the Schrödinger equation, i.e. a particle in free space where there is no potential present. This was the case that Schrödinger used to deduce the form of the wave equation for quantum particles. Let us now reconsider this system starting with the Schrödinger equation. Setting $V(x) = 0$, we can rewrite (3.22) as,

$$\frac{\partial^2 \psi(x)}{\partial x^2} = -k^2 \psi(x), \tag{3.24}$$

where $k = \sqrt{2ME}/\hbar$ is the wave number of the particle (since $E = \hbar^2 k^2/2M$). The solution to this is $\psi(x) = A\exp(\pm ikx)$, where A is the normalisation. The full time-dependent solution is, therefore,

$$\Psi(x,t) = Ae^{i(\pm kx - Et/\hbar)} = Ae^{i(\pm kx - \omega t)}, \tag{3.25}$$

where the last step follows because $E = \hbar\omega$ for a free particle. This is just a plane wave solution (travelling in either the $+x$ or $-x$ direction) and we note, in particular, that the probability of finding a particle anywhere does not depend on x, $|\Psi(x,t)|^2 = |A|^2$. In other words, the particle is equally likely to be found anywhere, as we might expect in free space.

In the next chapter, we will consider solving the Schrödinger equation in situations when the potential, $V(x)$ is not zero. This will lead to a number of interesting consequences and applications. Before we do that, we round off this chapter by introducing a few more key ideas in the theory of quantum mechanics.

3.5. Observables and Operators

Heisenberg understood that the main difference between classical and quantum mechanics is that observables in classical mechanics commute, while in quantum theory they do not. Classically, if we measure a position of a particle, and then its momentum, we would get the same numbers if we did it the other way round. This means that observables in classical physics can be represented simply by numbers (since 4 times 8 is the same as 8 times 4, i.e. numbers commute).

However, the same is not true in quantum mechanics and so observables in this case cannot just be numbers. Heisenberg used matrices for this purpose as he knew that they do not commute. In fact he developed a whole new formalism for quantum mechanics known as matrix mechanics (Hamilton had handily invented matrices some 60 years earlier). This is also the reason why a lot of quantum mechanics involves techniques from linear algebra and why a lot of the mathematics taught in an undergraduate physics degree involves manipulating matrices. But we will not use matrices just yet, we will meet them again later in the book. This is because here we are dealing with continuous systems and matrices could be awkward.

Let us first introduce the idea of operators. An operator is in general something that acts on one function to give us another one. Special kinds of operators (called Hermitian) represent observables. We will not spend too much time on this topic here as it will be covered in more detail later in the book, but it is useful to introduce the idea and see them in action. Here we will instead use differentials rather than matrices to represent operators. These are an equally good formalism to use since they also do not commute. We can see this in the following example. In one dimension, the position operator, $\hat{x}$, is simply x, when written in the position representation. In the same representation, the momentum operator, $\hat{p}$, representing the momentum observable, p, is given by

$$\hat{p} = -i\hbar \frac{\partial}{\partial x}. \tag{3.26}$$

We did not just pull this one out of a hat. This really is the only reasonable choice for momentum. Von Neumann showed this a long time ago and we will show why this must be the case later in the book. In the meantime, however, you will just have to take it on faith. We now show that position and momentum do not commute. Namely $\hat{x}\hat{p} \neq \hat{p}\hat{x}$, which can be seen by its operation on some wave function, $\psi(x)$,

$$x\left(-i\hbar\frac{\partial}{\partial x}\right)\psi(x) \neq \left(-i\hbar\frac{\partial}{\partial x}\right)x\psi(x). \tag{3.27}$$

This is because the right hand side is

$$-i\hbar\frac{\partial}{\partial x}x\psi(x) = -i\hbar\left(\psi(x) + x\frac{\partial}{\partial x}\psi(x)\right) = -i\hbar\left(1 + x\frac{\partial}{\partial x}\right)\psi(x), \tag{3.28}$$

which is not the same as the left hand side.

Another important concept in quantum mechanics is the eigenvalue equation. It has the form

$$\hat{Q}u(x) = qu(x) \tag{3.29}$$

Where $\hat{Q}$ is an operator, $u(x)$ is its eigenfunction and q is the corresponding eigenvalue. Let us consider the example of a free particle and the operator,

$$\hat{H} = -\frac{\hbar^2}{2M}\frac{d^2}{dx^2}. \tag{3.30}$$

From Eq. (3.22) with $V(x) = 0$, we see that we have

$$\hat{H}u(x) = Eu(x) \tag{3.31}$$

where E is the energy of the particle. This means that $\hat{H}$ is the operator for the energy of the free particle. We have already seen another form of an operator for the total energy of a system, $i\hbar\partial/\partial t$, which means we can write,

$$\hat{H}u(x) = i\hbar\frac{\partial}{\partial t}u(x). \tag{3.32}$$

This is an alternative (shorthand) way of writing the Schrödinger equation. The meaning of the eigenvalue equation is the following. If the physical system is in an eigenfunction of an operator, then upon measurement of that operator, we will obtain the corresponding eigenvalue with certainty.

3.6. The Superposition Principle

The superposition principle is the cornerstone of quantum mechanics and its failure would signal the end of quantum theory. It really is the key feature that distinguishes quantum physics from classical theories[9]. It states that whenever we have two solutions of the Schrödinger equation, i.e. two possible physical states of a system, then any superposition of the two is also a solution. This is very easy to show from the form of the Schrödinger equation.

The implications of this can be very bizarre. If an atom can be located here, this is one possible state, and if it could be located on the Moon, that is a second possible state, then it can also be simultaneously here and on the Moon. Or, if a photon can be blue or yellow, then it can also be both blue and yellow at the same time.

When we have a superposition of two different energies, for example, then the resulting state does not have a well defined energy. Let us show that in a more formal way now. Suppose we have an operator, $\hat{A}$ with two eigenstates ψ_1 and ψ_2 with corresponding eigenvalues a_1 and a_2, i.e.

$$\hat{A}\psi_1 = a_1\psi_1 \tag{3.33}$$

$$\hat{A}\psi_2 = a_2\psi_2. \tag{3.34}$$

What happens when we superpose the eigenstates? We obtain

$$\hat{A}(\psi_1 + \psi_2) = \hat{A}\psi_1 + \hat{A}\psi_2 = a_1\psi_1 + a_2\psi_2 \neq a(\psi_1 + \psi_2) \tag{3.35}$$

for any a. Therefore, if ψ_1 and ψ_2 are both eigenstates of the operator A with different eigenvalues a_1 and a_2, then their superposition is not. This fact is reflected in the statistical character of measurement outcomes in quantum mechanics.

Let us explain this a bit more. When the system is in a eigenstate of some observable, when we measure this observable, the outcome will (with 100% probability) be the eigenvalue corresponding to that eigenstate. Let us take position as an example. Say you prepare an electron at a specific place. If there are no forces disturbing the electron, next time you measure it you will find it in the same place. But what happens if you prepare an electron in two different places at the same time? Then, when you measure its position, it will sometimes be in one place and sometimes in the other one and you won't be able to predict which will happen at any given time. This randomness at the heart of nature is something that is very difficult to accept and has been the

[9]Richard Feynman famously said that "superposition is the only mystery in quantum mechanics". It is, however, a very big mystery!

source of much discussion and debate. The same goes for any quantum system, not just electrons, and for the measurement of any observable, not just position.

3.7. Expectation Values

In Section 3.2 we saw how the wave function can be used to determine the probability of finding a particle at a particular location. It can also be used to determine the expectation value of measuring some observable of the particle, e.g. position, momentum, or energy. It is straightforward to see how this works in the case of position. We know that the probability density for finding a particle at location x is $|\psi(x)|^2$, and so the expectation value for x is simply,

$$\langle x \rangle = \int_{-\infty}^{\infty} x\, |\psi(x)|^2\, dx.$$

This is usually written in the equivalent form,

$$\langle x \rangle = \int_{-\infty}^{\infty} \psi^*(x)\, x\, \psi(x)\, dx.$$

More generally, the expectation value for some observable O is given by

$$\langle O \rangle = \int_{-\infty}^{\infty} \psi^*(x)\, \hat{O}\, \psi(x)\, dx,$$

where $\hat{O}$ is the *operator* for the observable O. For position, the operator is simply x, for momentum it is $-i\hbar\frac{\partial}{\partial x}$, and for energy the operator is $i\hbar\frac{\partial}{\partial t}$, as we have seen. The expectation value for momentum, for example, is found by calculating,

$$\langle p \rangle = -i\hbar \int_{-\infty}^{\infty} \psi^*(x)\, \frac{\partial}{\partial x}\, \psi(x)\, dx.$$

One important property of the Schrödinger equation is that it is linear. This means that if we find two or more solutions, then any linear combination of those solutions will also be a solution. This introduces the idea of basis wave functions: the general solution to the Schrödinger equation can be written as a sum of components in different basis wave functions. This is similar to the idea that a general vector can be written as a sum of components in different basis directions, e.g. the x, y, and z directions in Cartesian coordinates. Often it is convenient to choose basis functions that are orthogonal, i.e. there is no overlap between them. The overlap between two wave functions $\psi(x)$ and $\phi(x)$ is,

$$\int_{-\infty}^{\infty} \phi^*(x)\, \psi(x)\, dx,$$

and the two states are orthogonal if this overlap is zero.

3.8. The Uncertainty Principle

Let us now exemplify the calculation of expectation values with the position and momentum observables. In the process we will demonstrate a very important result in quantum physics – Heisenberg's uncertainty principle.

For the sake of simplicity, let us choose a Gaussian wave function for the particle

$$\Psi(x) = \frac{1}{(2\pi\sigma^2)^{1/4}} e^{-x^2/4\sigma^2}, \tag{3.36}$$

where σ is a constant that represents the standard deviation of the Gaussian. The position variance of this state is $(\Delta x)^2 = \langle x^2 \rangle - \langle x \rangle^2$. The expectation value of x is zero, i.e. $\langle x \rangle = 0$, so

$$(\Delta x)^2 = \langle x^2 \rangle = \frac{1}{(2\pi\sigma^2)^{1/2}} \int_{-\infty}^{\infty} x^2 e^{-x^2/2\sigma^2} dx = \sigma^2. \tag{3.37}$$

Similarly, the momentum variance is $(\Delta p)^2 = \langle p^2 \rangle - \langle p \rangle^2$ and $\langle p \rangle = 0$, so

$$(\Delta p)^2 = \langle p^2 \rangle = \frac{1}{(2\pi\sigma^2)^{1/2}} \int_{-\infty}^{\infty} -\hbar^2 e^{-x^2/2\sigma^2} \frac{d^2}{dx^2} e^{-x^2/2\sigma} dx = \frac{\hbar^2}{4\sigma^2}. \tag{3.38}$$

The product of the uncertainties is, therefore,

$$\Delta x \Delta p = \frac{\hbar}{2}. \tag{3.39}$$

This conforms to the uncertainty principle for position and momentum, which states,

$$\Delta x \Delta p \geq \frac{\hbar}{2}. \tag{3.40}$$

In other words, the better we known x, i.e. the smaller Δx, the worse we know the momentum, i.e. the larger Δp. The two are exactly inversely proportional. We will revisit the uncertainty principle in a more formal (and more general) way later in the book.

3.9. Conceptual Foundations of Quantum Mechanics

Before we go on to describe how quantum mechanics can be applied to a range of different systems in the next chapter, it is worth pausing to think about what it all means. This question was asked right at the very beginning of quantum theory and it has driven a great deal of research in the past hundred years or so. Though we are still not quite sure what the answer is (despite the remarkable success of the theory), several distinct interpretations have emerged.

One of the ways you can start to ask questions about the meaning of quantum mechanics is to ask "what is it that waves in the wave function?". Born's answer would, of course, be that it is the probability amplitude waves. If pushed to explain, then he would say that the modulus square of the wave function is the probability to find the system in a given location and this is what waves. But, is this probability a property of the system that we observe or is it all in our heads. In other words, is the system waving or is our knowledge about the system waving, or is it the relationship between the system and myself that is waving? No one really has any answer to this, but various attempts to answer it have led to a number of different ways of looking at quantum mechanics.

The fact that objects in quantum mechanics can exist in many different states at the same time is difficult to reconcile with our everyday (classical) intuition. Dealing with this notion has led to a number of different interpretations of the meaning of quantum mechanics. Here we will mention some of the most important ones. Although they are different, they are not necessarily unrelated to each other.

The earliest one is the Copenhagen interpretation and was mainly constructed by Bohr himself. His take on this was basically as follows. Quantum mechanics is weird and implies some fundamental complementarity between properties in the everyday world. Namely, either we can set up our apparatus to find out one property or we can find some other property with a different arrangement, but the two cannot be found at the same time. Bohr called this principle the principle of complementarity and he believed in its widespread validity. Therefore, the key issue for Bohr was that the experimenal question – the measurement – defined what can be done. The measurement reveals physical properties and some experiments are simply mutually exclusive. Heisenberg, who was greatly influenced by Bohr, also subscribed to this view. Superpositions in this view are just our way of dealing with the weirdness of nature, but we don't really know and cannot know what the microscopic world is really like. In other words, we cannot ask questions about what is happening when we are not looking. Quantum physics is simply a tool for predicting measurement outcomes.

At the opposite extreme in some sense is the Many Worlds interpretation. This was actually first conceived by John von Neumann, but was fully developed into a framework by Hugh Everett in the 1950s. Everett did not like the fact that measurement plays such a central role, because its outcome is in general random. So Everett realised that the whole of quantum theory can be explained without the need for any measurements. The price to pay, however, is that we need to accept that we live in a multiverse and not in a universe. Everything that can happen does happen. Positions and momenta are known, but they exist in different universes and only one of these will be accessible to us. This view is appealing for many reasons, but it also leads to some bizarre consequences.

Finally, perhaps quantum mechanics simply fails at some level. Maybe some things just cannot be superposed. Different people have come up with variations of this interpretation and they all come under the collective term 'collapse of the wave function'. There are some big proponents of this idea with different ideas of how the collapse takes place. Unfortunately no experiment so far has ever indicated any failure of quantum mechanics. This is not to say, of course,

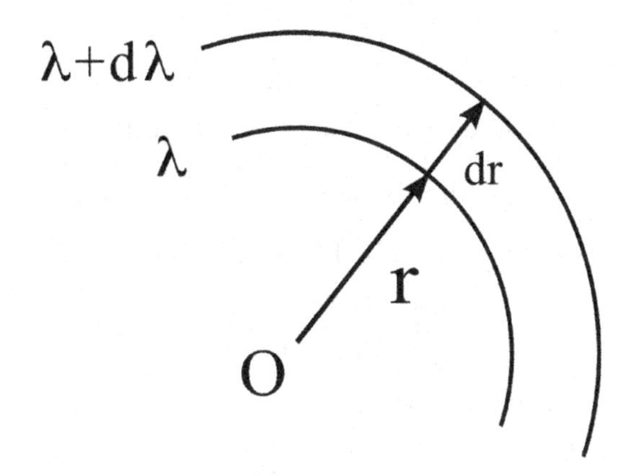

Fig. 3.1. How quantum mechanics reduces to classical. Consider an electron in a potential, V. At the radial distance r, from the origin, O, the electron has wavelength λ and at some infinitesimally larger distance $r + dr$, the wavelength is $\lambda + d\lambda$.

that future ones will not. As you learn more about quantum mechanics, you too will develop intuition as to which interpretation you are most inclined to accept. Until an experiment is able to distinguish these different interpretations you are, of course, free to choose any one you want. It is very much a question of taste.

Let us close this chapter by discussing one very important aspect of the consistency of quantum mechanics. We would like to show how it is possible to derive Newton's laws of motion from quantum mechanics. Rather than looking at the general case, which is a bit involved, though not too difficult, we will consider the specific case of an electron moving in an electromagnetic potential, V. What we would like to derive is that the radius of trajectory is governed by the balance between the centripetal force with the force derived from the electromagnetic potential.

At some radial distance r from the origin, the electron has wavelength λ and at some larger distance, $r + dr$, the wavelength is $\lambda + d\lambda$ (see Fig. 3.1). Now, we know that some integral number, n, of wavelengths must fit into the circumference, C. This gives

$$2\pi r = n\lambda \tag{3.41}$$

$$2\pi(r + dr) = n(\lambda + d\lambda). \tag{3.42}$$

Taking the ratio of these two equations, we get

$$\frac{r + dr}{r} = \frac{\lambda + d\lambda}{\lambda} \tag{3.43}$$

$$\Longrightarrow \frac{dr}{r} = \frac{d\lambda}{\lambda}. \tag{3.44}$$

From this we can obtain

$$\frac{1}{r} = -\frac{1}{k}\frac{dk}{dr}, \tag{3.45}$$

where $k = 2\pi/\lambda$ is the wavenumber. Now, we know that k can be expressed as

$$k = \sqrt{2M(E - V(r))/\hbar} \tag{3.46}$$

since the kinetic energy is $\hbar^2 k^2/2M$. Plugging this into (3.45) we get

$$\frac{1}{r} = \frac{dV/dr}{2(E - V(r))}. \tag{3.47}$$

Noticing that $Mv^2 = 2(E - V(r))$, we finally reach the following expression

$$\frac{Mv^2}{r} = \frac{dV}{dr}, \tag{3.48}$$

which exactly encapsulates Newtonian dynamics. It states that the centripetal force, on the left hand side, matches the force due to the electromagnetic potential, on the right hand side. Therefore, Newtonian physics is just a special case of quantum mechanics.

3.10. Exercises

1. Suppose two travelling plane waves of momentum $\hbar k$, but travelling in opposite directions, are superposed.

   Calculate the probability density of the resulting standing wave.

   What are the nodes of this state?

   Where is the probability maximal?

   The quantum formula for probability current is given by

   $$j = \frac{i\hbar}{2M}\left(\psi\frac{d}{dx}\psi^* - \psi^*\frac{d}{dx}\psi\right).$$

   Calculate this for the above superposition of two travelling opposite waves. How would you interpret your result?

2. The state of a particle is represented by the wave function

   $$\Psi(x) = A(a^2 - x^2) \quad -a \leq x \leq a$$

   and is zero outside of this region.

   Determine A from the normalisation condition.

   Calculate the expectation values of x and x^2.

   Calculate the expectation values of p and p^2.

   Check that the uncertainty principle holds. What would happen to the momentum if the region halved in size, i.e. if a was reduced by a factor of 2?

3. Show that the probability density $|\Psi(x,t)|^2 dx$ for a free particle (plane wave) is conserved with time.

   If, on the other hand, we have an unstable particle, with a lifetime τ (the mean disintegration time), the probability is no longer constant in time but behaves as

   $$P(t) = \int_0^L |\Psi(x,t)|^2 dx = e^{-t/\tau},$$

   where we assume that the particle exists on the interval $0 \leq x \leq L$.

   Suppose that the particle is initially a free particle in the state

   $$\Psi(x,0) = Ae^{i(kx-\omega t)}$$

   where A is normalisation.

   Derive the dependence of A on time if the probability is to behave as described.

What term V should be added in the Hamiltonian

$$H = \frac{-\hbar^2}{2M}\frac{d^2}{dx^2} + V$$

so that we recover the probability decrease above?

Now, suppose that the decay of the particle results in two photons, of identical frequency. Estimate their wavelengths as a function of τ if the whole process of decay is assumed to be energy conserving. You may fully neglect the kinetic part of the energy.

4. If $\psi_1(x,t)$ and $\psi_2(x,t)$ are both solutions to the 1-d time-dependent Schrödinger equation, show that the linear combination $\Psi = c_1\psi_1(x,t) + c_2\psi_2(x,t)$ is also a solution, where c_1 and c_2 are constants.

Chapter 4

Applications of Quantum Mechanics

We have spent quite a bit of time discussing the foundations and origins of the theory of quantum mechanics. Now seems to be a good time to see how it can be applied to a range of different systems and to see what it really says about the world. Although we have only considered quantum mechanics in one dimension, there are many important and beautiful applications of quantum mechanics that can be understood in this regime. In this chapter we review some of these.

4.1. Infinite Square Well

Apart from a free particle, possibly the simplest problem we can apply quantum mechanics to is a particle confined in an impenetrable trap. Suppose we have a particle confined in a potential of the form,

$$V(x) = \begin{cases} 0 & \text{for } 0 \leq x \leq L \\ \infty & \text{otherwise.} \end{cases} \tag{4.1}$$

This is essentially a one-dimensional box with length L and infinitely hard walls. What are the possible states of the particle?

First we will argue in an indirect but intuitive way what the states of this particle must be. This is a quantum particle, so according to de Broglie it must have a wavelength. The wavelength has to have nodes on the edges of the box, because the particle cannot exist outside of it. Therefore, the condition is

$$n\lambda_n = 2L \tag{4.2}$$

and this gives us an infinite set of discrete wavelengths. What is the energy of the particle? Well, we only have the kinetic energy, whose formula is

$$E_n = \frac{\hbar^2 k_n^2}{2M} = \frac{\pi^2 \hbar^2}{2ML^2} n^2. \tag{4.3}$$

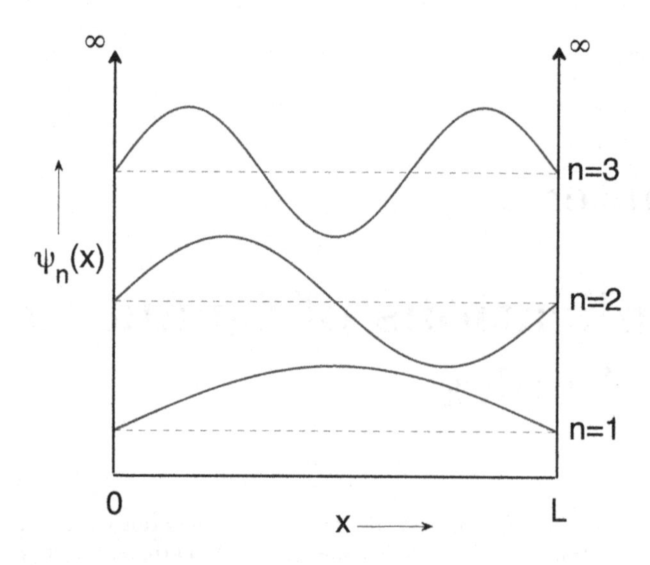

Fig. 4.1. Wave functions of the first three energy levels for a particle in an infinite square well.

The energy, therefore, grows as the square of the quantum number n.

We will now formally solve the Schrödinger equation just to show you that we get the same result. The time-independent Schrödinger equation can be solved in this case by dividing space up into regions where the potential is zero or infinite and solving these cases separately.

Outside the box, $V(x) = \infty$ and the wave function vanishes since there is no possibility of the particle being found there, i.e. $\psi(x) = 0$ for $x \notin [0, L]$. Inside the box, $V(x) = 0$ and we have the free particle case discussed above. The solutions in the different regions can be combined by ensuring that the wave function $\psi(x)$ is 'well-behaved' over all space. In particular, we need $\psi(x)$ to be everywhere finite, single-valued, continuous and smooth[1].

This means that we need to ensure that the plane-wave solutions inside the box continuously match the solutions outside it, i.e. $\psi(0) = \psi(L) = 0$. These are called boundary conditions and restrict the solutions that are allowed. The plane-wave solutions inside the box are $e^{ik_n x}$ and $e^{-ik_n x}$ or, equivalently, $\cos(k_n x)$ and $\sin(k_n x)$. The most general solution inside the box is therefore[2]

$$\psi(x) = A\sin(k_n x) + B\cos(k_n x) \qquad \text{for } x \in [0, L], \tag{4.4}$$

[1]This last condition of smoothness means that $d\psi/dx$ must be continuous wherever $V(x)$ is finite.

[2]Recall from Section 3.7 that a consequence of the fact that the Schrödinger equation is linear is that a linear superposition of any two or more solutions will also be a solution.

where A and B are constants. The conditions at the boundaries give,

$$\psi(0) = B = 0 \tag{4.5}$$
$$\psi(L) = A\sin(k_n L) = 0, \tag{4.6}$$

where we have made use of the result $B = 0$ in the second line. The second condition restricts what values of k_n are allowed. In particular, we need $k_n L = n\pi$, where n is an integer. In other words, the allowed wave functions inside the box are,

$$\psi_n(x) = A\sin\left(\frac{n\pi x}{L}\right) \qquad n = 1, 2, 3, \cdots \tag{4.7}$$

with corresponding *quantised* energies,

$$E_n = \frac{\hbar^2 k_n^2}{2M} = \frac{\pi^2\hbar^2}{2ML^2}n^2. \tag{4.8}$$

This is exactly what we found before with our intuitive argument.

The constant A can be found by normalisation,

$$1 = \int_{-\infty}^{\infty} |\psi_n(x)|^2\, dx = |A|^2 \int_0^L \sin^2\left(\frac{n\pi x}{L}\right) dx = |A|^2 L/2, \tag{4.9}$$

which gives, $A = \sqrt{2/L}$. The wave functions for different values of n are shown in Fig. 4.1.

There are many physical examples for which the infinite square well is a surprisingly accurate and useful model. Let us, for example, try to calculate the lowest energy level for an electron in a hydrogen atom. Imagine that the electron is confined in a box of size 10^{-10}m, which is a typical atomic size. The energy in the lowest level is then

$$E_1 \approx \frac{10^{-68}}{10^{-30}10^{-20}} = 10^{-18}\text{J} \tag{4.10}$$

This is roughly 10eV, and the real energy is 13.6eV, a remarkably close estimate.

What happens to the energy of the particle if the size of the box gets smaller? Then the energy goes up with inverse square of that size. Therefore, the more you try to confine quantum particles, the more energetic they get as they resist confinement. The floor you stand on is made up of atoms, and each of these has a nucleus and a bunch of electrons whizzing around. But most of the space in the atom is empty, since the nucleus and electrons are very small compared to the size of the atom. How is it that you do not fall through the floor? The usual answer is that the force of electromagnetic repulsion keeps you on the floor. But, in fact, it is the power of the uncertainty principle.

Let us now compare our quantum calculation of a particle in an infinite square box with the classical result. What would be the classical distribution for a particle inside a box. Well, if we do not have any information about the location of the particle, then it is equally likely to be anywhere. The probability density is, therefore, $1/L$ everywhere inside and, of course, zero outside. This is very different to any quantum mechanical eigenstate that we computed above. But, what about the quantum probability to find the particle on the left side of the box? It is:

$$\int_0^{L/2} \frac{2}{L}\sin^2(n\pi x/L) = \frac{1}{2}. \tag{4.11}$$

Of course, the same is also true for the probability that the particle is on the right side of the box. And this is the same as the classical particle. So, if we average over quantum probabilities, we obtain the classical ones.

4.2. The Quantum Harmonic Oscillator

Another important type of potential that is worth studying is the quantum harmonic oscillator. Just as the harmonic oscillator is important in classical physics as a prototype for more complex oscillations, the quantum harmonic oscillator plays a key role in quantum physics. We will see this later in the book particularly when we come to discuss quantum field theory in Chapter 10.

The classical harmonic potential describes systems, such as a mass on a spring, obeying Hooke's law or a marble rolling around in a parabolic container. The harmonic potential is given by,

$$V(x) = \frac{1}{2}M\omega^2x^2, \tag{4.12}$$

where ω is a constant describing the frequency of the potential, i.e. the rate at which a classical particle would complete a full oscillation. We know that to describe the motion of a quantum particle, we need to solve the Schrödinger equation. Substituting this potential into the one-dimensional Schrödinger equation, we get,

$$\frac{d^2\psi}{dx^2} = -\frac{2M}{\hbar^2}\left(E - \frac{1}{2}M\omega^2x^2\right)\psi(x). \tag{4.13}$$

The first three solutions to this are[3],

$$\psi_0(x) = \left(\frac{1}{\sqrt{\pi}b}\right)^{1/2} e^{-x^2/2b^2} \tag{4.14}$$

$$\psi_1(x) = \left(\frac{2}{\sqrt{\pi}b^3}\right)^{1/2} xe^{-x^2/2b^2} \tag{4.15}$$

$$\psi_2(x) = \left(\frac{1}{2\sqrt{\pi}b}\right)^{1/2}\left(\frac{2x^2}{b^2} - 1\right)e^{-x^2/2b^2}, \tag{4.16}$$

[3]For now, we just state the solutions, but you should check that they are correct by substituting back into the wave equation. When we come to discuss creation and annihilation operators in Section 8.13 we will see an elegant way of determining these wave functions.

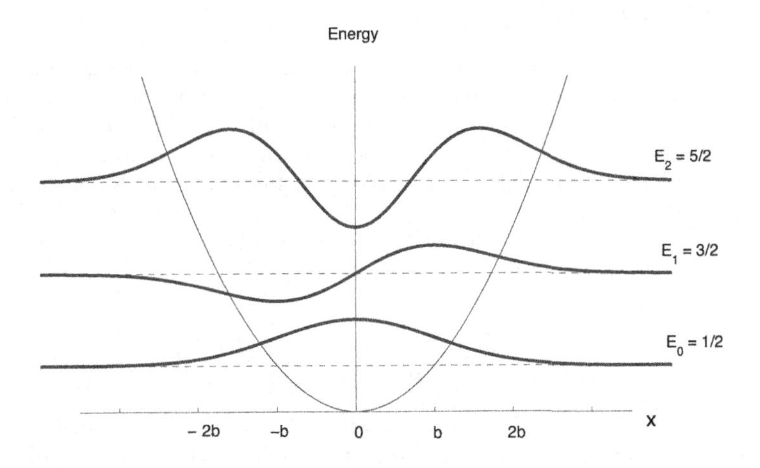

Fig. 4.2. The first three energy levels (in units of $\hbar\omega$) and wave functions of a quantum harmonic oscillator.

where $b = \sqrt{\hbar/(M\omega)}$ has the dimensions of length and corresponds to the classical turning point of an oscillator in the $n = 0$ ground state[4].

As expected, stationary states of a quantum harmonic oscillator exist only for certain discrete energy levels. The allowed energies are given by,

$$E_n = \left(n + \frac{1}{2}\right)\hbar\omega \qquad\qquad n = 0, 1, 2, ... \tag{4.17}$$

Figure 4.2 shows the first three energy levels and wave functions of a quantum harmonic oscillator. The energy levels have the special property that they are equally spaced by $\Delta E = \hbar\omega$. This result is different from a particle in a box, where the energy levels get increasingly further apart. We can also see in Fig. 4.2 that the wave functions extend well beyond the turning points into the classically forbidden region. This means that there is some probability that the particle will be found in regions where classical physics would never allow it to exist.

The quantum harmonic oscillator is widely used in quantum physics and you will see a lot more of it. The potential describing the bond between two particles, for example, is often well-approximated by a harmonic well and the very first quantum calculation essentially made use of a harmonic well when Planck described black body radiation using photons with quantised energies of $\hbar\omega$, i.e. the spacing of energy levels in a harmonic well.

[4]In other words, b is the maximum displacement in x that a particle with energy $E = (1/2)\hbar\omega$ can make in the potential $V = (1/2)M\omega^2x^2$. It is found simply by finding the value of x at which all the particle's energy has been converted to potential energy.

4.3. Tunnelling

We have seen in the case of a quantum harmonic oscillator that quantum theory allows a particle to be found in places where it could never exist according to classical physics. One fascinating consequence of this is that it is possible for a particle to simply tunnel through a potential barrier that it doesn't have enough energy to pass over. This is one of the most amazing examples of the power of quantum mechanics.

We can see how this works by considering the example of a rectangular barrier corresponding to the potential,

$$V(x) = \begin{cases} 0 & x < 0 \\ U_0 & 0 \le x \le L \\ 0 & x > L \end{cases} \tag{4.18}$$

as shown in Fig. 4.3, and a particle with total energy $E < U_0$ incident on the barrier from the left. Classically, it is impossible for the particle to pass to the right hand side of the potential. Let us now see what happens if we treat the particle quantum mechanically.

We can treat the three regions of the potential (4.18) separately. The solution to the Schrödinger equation is just a plane wave for $x < 0$ and $x > L$, i.e.

$$\psi(x) = \begin{cases} Ae^{ikx} + Be^{-ikx} & x < 0 \\ Ce^{ikx} + De^{-ikx} & x > L \end{cases} \tag{4.19}$$

where A, B, C and D are constants. The solutions are superpositions of plane waves travelling to the right and to the left. For $x < 0$, these correspond to the incident wave and the wave reflected from the barrier. For $x > L$, there is a transmitted wave, but nothing to cause a reflection and so we can set $D = 0$.

The region $0 \le x \le L$ is classically forbidden because $E < U_0$. The Schrödinger equation in this region can be rewritten as

$$\frac{d^2\psi(x)}{dx^2} = \alpha^2\psi(x), \tag{4.20}$$

where $\alpha^2 = \frac{2M(U_0 - E)}{\hbar^2} > 0$. This has solutions $\psi(x) = \exp(\pm\alpha x)$. This is not an oscillating wave but an exponentially decaying (or increasing) function. So the general solution in this region is,

$$\psi(x) = Fe^{\alpha x} + Ge^{-\alpha x} \qquad 0 \le x \le L \tag{4.21}$$

for constants F and G. We can find the values of the constants A, B, C, F and G (recall that $D = 0$) in equations (4.19) and (4.21) by applying the continuity conditions at the boundaries between the regions. This means that we need $\psi(x)$ and $d\psi/dx$ to be continuous at the boundaries. At $x = 0$, these conditions give

$$A + B = F + G \tag{4.22}$$

$$ik(A - B) = \alpha(F - G). \tag{4.23}$$

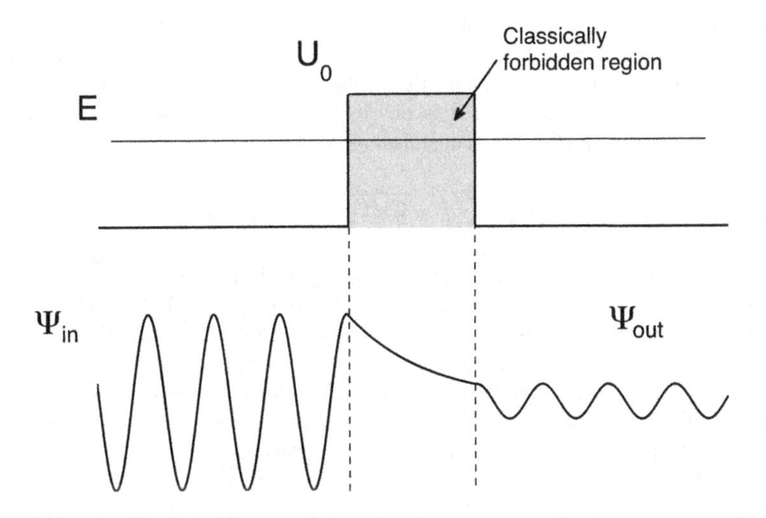

Fig. 4.3. Top: A particle with energy, E is incident from the left on a rectangular barrier of height $U_0 > E$. Bottom: The wave function for the particle as it tunnels through the barrier. The outgoing wave function has the same energy as the incoming one, but reduced amplitude because the probability of tunnelling is less than one.

At $x = L$, these two conditions give,

$$Ce^{ikL} = Fe^{\alpha L} + Ge^{-\alpha L} \tag{4.24}$$

$$ikCe^{ikL} = \alpha(Fe^{\alpha L} - Ge^{-\alpha L}). \tag{4.25}$$

Eliminating F and G from equations (4.22)–(4.25) and solving for the ratios B/A and C/A, we obtain the following results (the details are provided in the next section) for the reflection R and transmission T coefficients

$$R = \frac{|B|^2}{|A|^2} = \left(1 + \frac{4E(U_0 - E)}{U_0^2 \sinh^2(\alpha L)}\right)^{-1} \tag{4.26}$$

$$T = \frac{|C|^2}{|A|^2} = \left(1 + \frac{U_0^2 \sinh^2(\alpha L)}{4E(U_0 - E)}\right)^{-1}. \tag{4.27}$$

R and T are respectively the probabilities that the particle will be reflected or transmitted by the barrier. It is easy to check that $R + T = 1$ as it has to be because (even in the strange world of quantum mechanics) the particle can only ever pass through the barrier or be reflected. The remarkable feature of this result is that $T > 0$. In other words, there is some finite probability that the particle will pass straight through the barrier even though this is forbidden by classical physics.

In reality, of course, most potential barriers cannot be approximated by a rectangular barrier. So let us consider the case of a more general potential barrier, $V(x) > E$. This is a much more difficult problem to solve and beyond the scope of this book. However it is possible to find a very simple and useful approximate expression for the transmission coefficient

$$T \approx \exp\left(-\frac{2}{\hbar}\int \sqrt{2M(V(x)-E)}\,dx\right), \tag{4.28}$$

where the integral is taken over the classically forbidden region, i.e. where $E < V(x)$ and M is the mass of the tunnelling particle. This formula is very nearly correct and will serve our purposes of estimating rates and probabilities of tunnelling[5]. Strictly, it is an approximation in the limit that we have a high, wide barrier, i.e. the rate of tunnelling is small. Note that when we take the classical limit, $\hbar \to 0$, the probability for tunnelling goes to zero as it should, since we don't see tunnelling in our everyday classical world.

Let us now use our formula to estimate some tunnelling rates. First of all, imagine that $V(x) = V$ between the two points $x_2 - x_1 = d$. The probability of tunnelling is then

$$T = e^{-2d\sqrt{\frac{2M}{\hbar^2}(V-E)}}. \tag{4.29}$$

Let us put in some numbers. Suppose you weighed 50kg and were attempting to tunnel through a 30J potential barrier which was 1m wide, while walking at 1m/s[6]. Your kinetic energy would be $E = \frac{1}{2}Mv^2 = 25$J and so the probability of you tunnelling through the barrier would be approximately equal to:

$$T \approx e^{-2\times 1\times\sqrt{100(30-25)}/(1.05\times 10^{-34})} \approx e^{-4.3\times 10^{35}}, \tag{4.30}$$

which is a very small number indeed. So don't try this at home.

But for small particles, such as an electron, the numbers look much more promising. Suppose that an electron with energy 5.1eV approaches a barrier of height 6.8eV and a width of $L = 750$pm. What is the approximate probability that the electron will appear on the other side of the barrier? The answer is $T \approx 43 \times 10^{-6}$. Admittedly a small number, but for every million electrons,

[5]The exact formula (which includes the appropriate normalisation) is

$$T = \frac{e^{-2I}}{(1+\frac{1}{4}e^{-2I})^2}$$

where

$$I = \int_{x_1}^{x_2} dx k(x)$$

and

$$k(x) = \sqrt{\frac{2M}{\hbar^2}(V(x)-E)}.$$

[6]...or through the wall of Platform 9 3/4 at King's Cross Station in London.

about 43 will make it through, so it is not something we can ignore. What about a proton? Protons are almost two thousand times heavier than electrons and, since the mass of the particle appears in the exponential of the transmission probability, the answer is very sensitive to mass. In fact, we obtain $T \approx 10^{-420}$ – a very much smaller probability.

4.4. Reflection and Transmission Coefficients

In this section we show you the details of how the reflection and transmission coefficients given by (4.26) and (4.27) can be obtained from equations (4.22) – (4.25). This section is provided for completeness with the interested reader in mind. It can safely be skipped over by everyone else.

We start by using equations (4.22) and (4.23) to find F and G in terms of A and B, i.e. we take $\alpha \times$ (4.22) + (4.23) and $\alpha \times$ (4.22) – (4.23). This gives

$$\begin{aligned} 2\alpha F &= A(\alpha + ik) + B(\alpha - ik) \\ 2\alpha G &= A(\alpha - ik) + B(\alpha + ik). \end{aligned}$$

Substituting, (4.24) into (4.25), we get

$$\begin{aligned} ik(Fe^{\alpha L} + Ge^{-\alpha L}) &= \alpha(Fe^{\alpha L} - Ge^{-\alpha L}) \\ \Longrightarrow Fe^{\alpha L}(\alpha - ik) &= Ge^{-\alpha L}(\alpha + ik). \end{aligned}$$

Then using the expressions for F and G, we get

$$(A(\alpha + ik) + B(\alpha - ik))e^{\alpha L}(\alpha - ik) = (A(\alpha - ik) + B(\alpha + ik))e^{-\alpha L}(\alpha + ik).$$

Dividing through by B and rearranging gives

$$\begin{aligned} \frac{A}{B}\left[e^{\alpha L}(\alpha^2 + k^2) - e^{-\alpha L}(\alpha^2 + k^2)\right] &= -e^{\alpha L}(\alpha - ik)^2 + e^{-\alpha L}(\alpha + ik)^2 \\ &= -2(\alpha^2 + k^2)\sinh(\alpha L) + 4i\alpha k\cosh(\alpha L). \end{aligned}$$

Taking the modulus squared of both sides

$$\begin{aligned} \frac{|A|^2}{|B|^2}(\alpha^2 + k^2)^2 4\sinh^2(\alpha L) &= (\alpha^2 - k^2)^2 4\sinh^2(\alpha L) + 16k^2\alpha^2\cosh^2(\alpha L) \\ &= (\alpha^2 - k^2)^2 4\sinh^2(\alpha L) + 16k^2\alpha^2(1 + \sinh^2(\alpha L)) \\ &= 4(\alpha^2 + k^2)^2\sinh^2(\alpha L) + 16k^2\alpha^2. \end{aligned}$$

Rearranging gives

$$\frac{|A|^2}{|B|^2} = 1 + \frac{4k^2\alpha^2}{(\alpha^2 + k^2)^2\sinh^2(\alpha L)}.$$

Finally, inverting gives

$$R = \frac{|B|^2}{|A|^2} = \left(1 + \frac{4k^2\alpha^2}{(\alpha^2 + k^2)^2\sinh^2(\alpha L)}\right)^{-1}.$$

The transmission coefficient $T = |C|^2/|A|^2$ can be calculated in a similar way or simply by making use of the result $T = 1 - R$.

4.5. Tunnelling in Action

Tunnelling is not just some strange quirk of quantum theory, but a very real effect that can be seen and measured. There are even technologies that rely on it to work. One important process that can be explained by tunnelling is the decay of radiocative nuclei by the emission of alpha particles. In the early days of quantum mechanics, the alpha decay of radioactive nuclei was a perplexing problem. The reason was that the nuclear forces were seen to be too strong for any particle to be able to leave the nucleus. In fact, that is the whole reason why we have nuclei in the first place. They hold together because nuclear forces are far stronger than the electromagnetic repulsion. It seemed strange that alpha particles emitted from a wide range of sources all had very similar energies (typically a few MeV), yet the half-lives of these sources varied by many orders of magnitude.

George Gamow[7] was the first to explain alpha decay via tunnelling in 1928. He considered an alpha particle rattling around inside the nucleus with insufficient energy to overcome the potential barrier provided by the strong nuclear force. However, with some small probability, the alpha particle could tunnel through the barrier and escape. From equation (4.28), we see that the transmission probability depends exponentially on the difference between the particle's energy, E and the barrier height $V(x)$. This explains why small variations in the energies of emitted alpha particles can lead to very large differences in the rates of emission (or half-lives).

The potential seen by an alpha particle is shown in Fig. 4.4. Using (4.28), the transmission probability can be calculated as[8],

$$T(E) \approx \exp\left(-\frac{Ze^2}{\epsilon_0 \hbar}\sqrt{\frac{M_\alpha}{2E}} + \frac{4e}{\hbar}\sqrt{\frac{ZRM_\alpha}{\pi\epsilon_0}}\right), \tag{4.32}$$

where Z is the atomic number of the nucleus *after* the alpha decay, E is the energy of the emitted particle, R is the radius of the parent nucleus, M_α is the mass of the alpha particle, and ϵ_0 is the permittivity of free space ($8.85 \times 10^{-12}\,\mathrm{Fm}^{-1}$).

[7]George Gamow is known for many things in physics including his famous 1948 paper [5] *The Origin of Chemical Elements* authored by Alpher, Bethe, and Gamow (a deliberate pun on the first three letters of the Greek alphabet: α, β, γ; indeed, Alpher was asked to join the author list just because of this and he did no actual work for the paper). As part of this, he estimated the residual cosmic microwave background from the Big Bang, which was later experimentally confirmed by Penzias and Wilson in 1964. He also wrote the excellent series of *Mr Tompkins* books on physics and biology.

[8]The expression (4.32) can be calculated (as an exercise), by substituting the form of the potential, shown in Fig. 4.4, into (4.28) and making use of the identity

$$\int_R^b \sqrt{\frac{b}{r} - 1}\, dr = b\left[\cos^{-1}\sqrt{\frac{R}{b}} - \sqrt{\frac{R}{b} - \frac{R^2}{b^2}}\right]. \tag{4.31}$$

This can be simplified further by making the wide-barrier approximation, $R \ll b$, which gives $\cos^{-1}\sqrt{R/b} \approx \pi/2 - \sqrt{R/b}$.

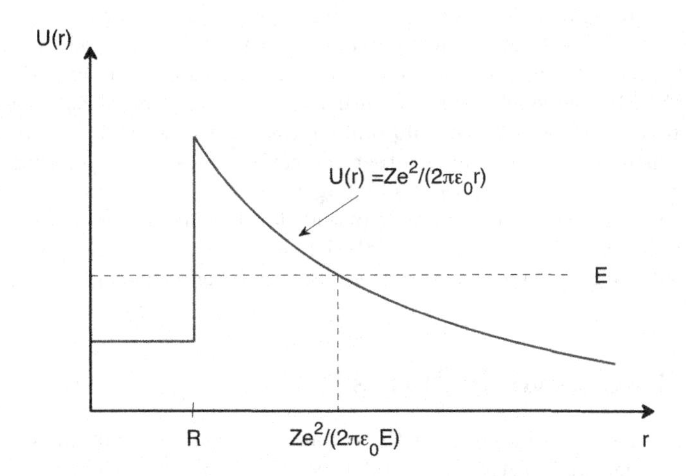

Fig. 4.4. Potential experienced by an alpha particle with energy E. Inside the nucleus ($r < R$), the particle experiences a constant potential. If the particle tunnels through the barrier, it experiences Coulomb repulsion outside the nucleus since the alpha particle and the nucleus are both positively charged.

To turn the probability $T(E)$ into a decay rate, we must multiply it by the rate at which the alpha particle makes collisions with the potential barrier. Typically, there are about 10^{21} collisions per second and so the decay rate is $\lambda \approx 10^{21}T(E)$. The half-life can then be calculated from this decay rate using

$$t_{1/2} = \frac{\ln 2}{\lambda}. \tag{4.33}$$

One technological application of quantum mechanical tunnelling is the scanning tunnelling microscope (STM). This is a non-optical microscope which employs the principles of quantum mechanics to image surfaces on an atomic scale[9]. It works by scanning an atomically sharp probe (the tip) over a material within a few nanometers of the surface. A voltage is applied between probe and the surface and electrons will tunnel from the tip to the surface (or vice-versa depending on the polarity), resulting in a weak electric current. The size of this current is exponentially dependent on the distance between probe and the surface. For a current to occur the substrate being scanned must be conductive, which means that electrical insulators cannot be scanned using an STM.

A servo loop keeps the tunnelling current constant by adjusting the distance between the tip and the surface. This adjustment is done by placing a voltage

[9]This feat is all the more remarkable when we consider just how small an atom is. It is sometimes said that there are more atoms in a single grain of sand than there are grains of sand on all the beaches in the world. A simple calculation with reasonable assumptions shows that this really is true.

on the electrodes of a piezoelectric element. By scanning the tip over the surface and measuring the height, which is related to the voltage applied to the piezo element, one can thus reconstruct the surface structure of the material under study. STMs are widely used in both industrial and fundamental research. They provide a three-dimensional profile of the surface, which is very useful for characterising surface roughness, observing surface defects and determining the size and conformation of molecules and aggregates on the surface. Examples of advanced research using the STM are provided by current studies in the Electron Physics Group at NIST and at the IBM Laboratories. Several other recently developed scanning microscopies also use the scanning technology developed for the STM.

4.6. Two Level Systems

We would now like to talk about the dynamics of coupled quantum states and how it can be described within our formalism. The simplest example we can study involves two levels. Imagine that a particle is sitting in one quantum level, described by the wave function ψ_1 and that we would like to excite the particle to another state described by ψ_2. We can do this, for example, by giving the particle enough energy to make the transition between the two levels. How is this situation handled quantum mechanically?

The key is to write down the correct Hamiltonian. It is

$$\hat{H} = \hbar\Omega(\hat{\sigma}_+ + \hat{\sigma}_-). \tag{4.34}$$

Here $\hat{\sigma}_+$ and $\hat{\sigma}_-$ are respectively the raising and lowering operators[10] defined by the following operations:

$$\hat{\sigma}_+\psi_1 = \psi_2 \quad \text{and} \quad \hat{\sigma}_-\psi_2 = \psi_1. \tag{4.35}$$

We now need to solve the Schrödinger equation

$$\hbar\Omega(\hat{\sigma}_+ + \hat{\sigma}_-)\psi(t) = i\hbar\frac{\partial\psi}{\partial t}. \tag{4.36}$$

We assume a trial solution of the form

$$\psi(t) = a_1(t)\psi_1 + a_2(t)\psi_2. \tag{4.37}$$

By plugging this into the above equation we obtain

$$\hbar\Omega(a_1(t)\psi_2 + a_2(t)\psi_1) = i\hbar\left(\frac{\partial a_1(t)}{\partial t}\psi_1 + \frac{\partial a_2(t)}{\partial t}\psi_2\right). \tag{4.38}$$

We now match terms that are proportional to ψ_1 and ψ_2 on both sides and get a system of coupled differential equations

$$\hbar\Omega a_1(t) = i\hbar\frac{\partial a_2(t)}{\partial t} \tag{4.39}$$

$$\hbar\Omega a_2(t) = i\hbar\frac{\partial a_1(t)}{\partial t}. \tag{4.40}$$

[10] We will meet them again in more detail in Section 8.13.

This is solved in the following way. We differentiate the second equation once more and substitute in the first. This leads to

$$\frac{\partial^2 a_1(t)}{\partial t^2} = -\Omega^2 a_1(t). \tag{4.41}$$

Let us further suppose that the system starts in the state ψ_1. Then the solution to the above is

$$a_1(t) = \sin \Omega t. \tag{4.42}$$

This means that the probability for the system to be in ψ_1 change with time as

$$|a_1(t)|^2 = \sin^2 \Omega t \tag{4.43}$$

and, therefore, oscillates in a sinusoidal fashion called Rabi flopping[11].

4.7. Cold Matter

Bose–Einstein condensation is a process whereby quantum particles obeying Bose statistics, reach a state where their identity becomes completely obscured. In this case, the particles all behave like a single, albeit large, quantum system. This impressive phenomenon, predicted by Bose and Einstein in the 1920s, was observed in dilute atomic gases only in 1995 and was awarded a Nobel prize in 2001. The long delay between its prediction and observation is because atomic condensates are extremely difficult to create. The problem is that quantum statistics can only be observed at very low temperatures – the lowest ever created[12]. This meant that innovative new cooling methods had to be devised[13]. One such method is called laser cooling and simply involves illuminating an atom with laser light of just the right frequency.

We have seen that a resonant interaction of light with an atom leads to absorption during which momentum is transferred to the atom. If the atom emits spontaneously, this then on average leads to no recoil as spontaneous emission is isotropic. Suppose that the beam of light is shone onto an atom moving in a certain direction. When a photon is absorbed, the atom slows down and the velocity reduction $\Delta\mathbf{v}$ is inferred from

$$M\Delta\mathbf{v} = \hbar\mathbf{k} \tag{4.44}$$

[11] A very similar effect is thought to account for the so-called solar neutrino problem. Theoretical models predict that a certain flux of electron neutrinos (ν_e) should be produced by the Sun, however Earth-based measurements only detect about one-third of the predicted value. The explanation is that the neutrinos quantum mechanically oscillate into two different 'flavours': the muon neutrino (ν_μ) and the tau neutrino (ν_τ), so on average only a third are found as electron neutrinos.

[12] ....and certainly colder than anything that occurs naturally in the Universe. Even deepest space is flooded with microwave background radiation left over from the Big Bang that corresponds to a temperature of about 2.7K.

[13] The 1997 Nobel prize in physics was awarded to Steven Chu, Claude Cohen-Tannoudji and William Phillips 'for the development of methods to cool and trap atoms with laser light'.

where M is atoms mass and $\mathbf{k}$ photon's wave vector. This means that the atom is cooled. Let us estimate how many photon absorptions we need to bring an atom to a standstill.

The root mean square velocity is given by

$$v = \sqrt{\frac{3kT}{M}}. \tag{4.45}$$

For rubidium-87, one of the most commonly used elements for creating condensates in the laboratory, at a temperature $T = 600\text{K}$, the root mean square velocity is $v \approx 400\text{m/s}$. It takes about

$$N = v/\Delta v \approx 70000 \tag{4.46}$$

cycles of emission and absorption to stop this atom (assuming a laser wavelength of $\lambda \approx 800\text{nm}$). However, the atom will never be stopped in reality. This is because the light, which is absorbed by the atom to slow it down, is then re-emitted in a random direction. Although, for all atoms, this recoil averages to zero, any given atom will undergo a form of random walk[14], which is equivalent to heating. The equilibrium temperature is found by balancing the cooling rate (due to photon absorption) with the heating rate (due to the random emission of photons). This gives[15]

$$kT = -\frac{\hbar\Gamma}{4}\left(\frac{\Gamma}{2\delta} + \frac{2\delta}{\Gamma}\right), \tag{4.47}$$

where Γ is the linewidth of the excited state of the atom, i.e. Γ^{-1} is the excited state's lifetime, and δ is the angular frequency of the detuning of the lasers from atomic resonance. We can find the minimum temperature by straightforward calculus. If we set the derivative of T with respect to δ equal to zero and then solve for δ, we can show that the temperature is a minimum when $\delta = -\Gamma/2$. The minus sign means that the frequency of the lasers must be lower than the resonant frequency. This is often called 'red detuning' because the frequency is detuned from the resonance towards the red, i.e. lower frequency, end of the spectrum. Substituting $\delta = -\Gamma/2$ into Eq. (4.47) we get an expression for the residual temperature

$$T = \frac{\hbar\Gamma}{2k}. \tag{4.48}$$

This is the so-called Doppler limit for cooling.

As an example, the Doppler limit for sodium atoms cooled on the resonance transition at 589 nm where $\Gamma/2\pi = 10\,\text{MHz}$, is $240\,\mu\text{K}$, and corresponds to an rms velocity of about $0.3\,\text{m/s}$. The temperature limits for other atoms are very

[14] This is like Brownian motion, but in the case of Brownian motion atoms in the surrounding gas provide kicks for a speck of dust to move around.

[15] For more details about cooling atoms we refer the reader to Foot C.J. (2005). *Atomic Physics*, Oxford [6] or Phillips W.D. (1998). Nobel Lecture: Laser cooling and trapping of neutral atoms. *Reviews of Modern Physics* **70**: 721–741 [7].

similar and are in the mK regime. This is extremely cold and the successful experimental demonstration of laser cooling has been a wonderful achievement. However, to obtain a clear condensate we need to go to a few orders of magnitude lower still. Let us perform a 'back of an envelope' calculation to estimate this temperature.

Suppose we want to condense N atoms contained within some volume V. The average distance between atoms is

$$d = \left(\frac{V}{N}\right)^{1/3}. \tag{4.49}$$

Condensation takes place when the de Broglie wavelength becomes comparable to the interparticle separation, i.e.

$$d \approx \lambda. \tag{4.50}$$

According to de Broglie's hypothesis the relationship between atom's velocity v and wavelength λ is

$$v = \frac{h}{M\lambda}. \tag{4.51}$$

This means our condition is

$$\frac{h}{Mv} \approx d. \tag{4.52}$$

We can then substitute in for v using the equipartition of energy

$$\frac{3}{2}kT = \frac{Mv^2}{2}. \tag{4.53}$$

This gives,

$$T = \frac{h^2}{3kM}\rho^{2/3}, \tag{4.54}$$

where $\rho = N/V$ is the number density. Taking the case of rubidium-87 and using a typical experimental number density of $\rho = 10^{14}\text{cm}^{-3}$, we get $T \approx 10^{-7}\text{K}$, which agrees very well with what is observed in experiments[16]. This means that to achieve condensation, we need to get to temperatures about four orders of magnitude lower than what can be achieved by laser cooling. Historically, this was achieved by first trapping laser cooled atoms and then cooling them further with a technique known as evaporative cooling. It is very similar to the way evaporation works in a cup of hot coffee. A cup of coffee is made up of many molecules flying around and bumping into each other. The temperature of the coffee is just a measure of the average energy that these fast-moving molecules have. From time to time two molecules will collide in such a way that one of them ends up with most of the energy, sometimes even gaining enough energy to fly out of the cup. Since these molecules are going fast compared to the rest of the molecules, they take with them more than their fair share of energy, and

[16]This number density is about 100000 times smaller than the density of the air that you are breathing. So these Bose–Einstein condensates really are very dilute gases.

the molecules that are left behind have less energy on average than they did before the fast molecules escaped. For every molecule that is kicked out the temperature of the coffee decreases a tiny amount. Remarkably, this technique (when applied to atomic gases) enables temperatures of a few nanoKelvin to be achieved and for Bose–Einstein condensation to be observed.

4.8. Exercises

1. This question addresses the issue of how we can recover some aspects of classical behaviour from quantum mechanics. Suppose an electron is trapped in the ground state of a one dimensional infinite square well.

   What is the probability of detecting the electron in the left one-third of the well?

   What is the probability of detecting the electron in the middle-third of the well?

   How do the above probabilities change if the electron is in a very excited state?

   What can you conclude about the quantum to classical transition limit?

2. Suppose that a particle in the infinite potential well is in a superposition of its two lowest states. Solve the time dependent Schrödinger equation to obtain its state at time t.

   What is the probability at time t of obtaining the ground state? What is the probability of obtaining the first excited state?

   How long does it take the particle to get back to its original state?

   Imagine now a classical particle whose kinetic energy is the difference between the two energies of the infinite square well.

   How long would it take this particle to take a round trip from one wall to another and back?

   How does this compare to the quantum period oscillations?

3. Which probability density represents a quantum harmonic oscillator with $E = \frac{5}{2}\hbar\omega$?

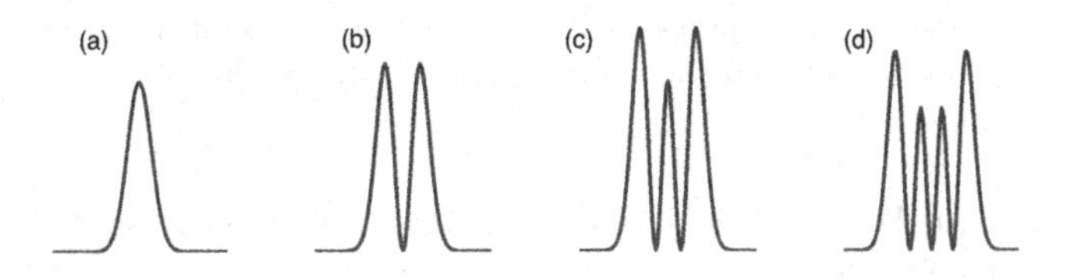

4. (a) Sketch the wave function for the ground state and the first two excited states of a particle in a harmonic potential.

   (b) Show that $b = \sqrt{\hbar/M\omega}$ corresponds to the classical turning point of a particle in the ground state of a harmonic potential.

(c) The Schrödinger equation describing a particle with energy E and mass M in a harmonic oscillator with frequency ω is,

$$-\frac{\hbar^2}{2M}\frac{d^2\psi(x)}{dx^2} = \left(E - \frac{1}{2}M\omega^2 x^2\right)\psi(x)$$

By substitution, show that

$$\psi(x) = A\exp\left(-\frac{M\omega x^2}{2\hbar}\right)$$

is a solution to (2) when $E = \frac{1}{2}\hbar\omega$. (Note that A is a constant).

5. (a) The wave function for the first excited state of an harmonic oscillator has form

$$\psi_1(x) = Axe^{-x^2/2b^2},$$

where $b = \sqrt{\hbar/(M\omega)}$. Find A so that the wave function is correctly normalised.
You may assume the following result: $\int_{-\infty}^{\infty} x^2 e^{-x^2}\, dx = \sqrt{\pi}/2$

(b) Show that the state $\psi_1(x)$ found in part (a) is orthogonal to the ground state wave function

$$\psi_0(x) = \left(\frac{1}{\sqrt{\pi}b}\right)^{1/2} e^{-x^2/2b^2}.$$

(c) The Hamiltonian for this harmonic oscillator is

$$H = -\frac{\hbar^2}{2M}\frac{d^2}{dx^2} + \frac{1}{2}M\omega^2 x^2.$$

Show that the state $\psi_1(x)$ found in part (a) is an eigenstate of this Hamiltonian and find the eigenvalue associated with it.

(d) What are the energy levels for a particle of mass, M, trapped in the potential

$$V(x) = \infty \qquad x < 0$$
$$V(x) = \frac{1}{2}M\omega^2 x^2 \qquad x \geq 0$$

A detailed calculation is not required. You may find it helpful to think in terms of the symmetry properties of the wave functions.

6. A particle is in the ground state of a one-dimensional infinite square well with walls at $x = 0$ and $x = L$. At $t = 0$, the right hand wall of the well is suddenly moved to $2L$ (while the left hand wall remains at $x = 0$). Find the probabilities that the particle will be found in the ground state and the first and second excited states of the expanded well.

7. Estimate the half-life for α-decay of thorium ($Z = 90$) given that the emitted α particles have energy 4.05 MeV and the radius of the nucleus is approximately 9×10^{-15} m. You may assume that the α particle makes about 10^{21} collisions with the potential barrier per second.

Chapter 5

Schrödinger Equation in Three-Dimensions

5.1. Three-Dimensional Box

Up until now we have only discussed the Schrödinger equation in one-dimension, e.g. a particle in an infinite square well. Of course, for most real-world situations, we need to be able to study the behaviour of particles in three-dimensions. In this chapter, we will consider how our formalism can be extended to deal with three-dimensional systems and will use it to study the structure of the hydrogen atom in depth.

The one-dimensional Schrödinger equation,

$$-\frac{\hbar^2}{2M}\frac{d^2\psi}{dx^2}+V(x)\psi=E\psi, \tag{5.1}$$

can be generalised to three-dimensions in a straightforward fashion. In Cartesian coordinates (i.e. x, y, z), it is simply

$$-\frac{\hbar^2}{2M}\left(\frac{\partial^2\Psi}{\partial x^2}+\frac{\partial^2\Psi}{\partial y^2}+\frac{\partial^2\Psi}{\partial z^2}\right)+V(x,y,z)\Psi=E\Psi. \tag{5.2}$$

Let us begin by considering the case of the infinite square box, which is the 3-d generalisation of the infinite square well and has the form

$$V(x,y,z)=\begin{cases}0 & \text{for } 0\leq x,y,z\leq L\\ \infty & \text{otherwise.}\end{cases} \tag{5.3}$$

Outside the box, the wave function is zero and inside the box, $V(x,y,z)=0$, and we can find the wave function by using the separation of variables technique we saw earlier. In particular, we pick a trial solution of the form, $\Psi(x,y,z)=\psi(x)\psi(y)\psi(z)$. Substituting into (5.2) and dividing through by $\Psi(x,y,z)$ we get

$$-\frac{\hbar^2}{2M}\left(\frac{1}{\psi(x)}\frac{d^2\psi(x)}{dx^2}+\frac{1}{\psi(y)}\frac{d^2\psi(y)}{dy^2}+\frac{1}{\psi(z)}\frac{d^2\psi(z)}{dz^2}\right)=E. \tag{5.4}$$

Each term on the left depends only on one variable and so each can be varied independently of the others. This means that (5.4) can be decomposed into three independent equations:

$$-\frac{\hbar^2}{2M}\frac{d^2\psi(x)}{dx^2} = E_x\psi(x)$$
$$-\frac{\hbar^2}{2M}\frac{d^2\psi(y)}{dy^2} = E_y\psi(y)$$
$$-\frac{\hbar^2}{2M}\frac{d^2\psi(z)}{dz^2} = E_z\psi(z),$$

where $E = E_x + E_y + E_z$. Each of these equations is the same as the equation for an infinite square well. That means that the wave function inside an infinite square box is simply the product of an infinite square well wave function for each of the three dimensions[1].

We would now like to study the structure of atoms in a fully quantum manner by solving the 3-d Schrödinger equation. Let us start with the hydrogen atom since it is the simplest atom with only a single electron. We have already talked about Bohr's model of the hydrogen atom and seen that, despite its success, it has a number of shortcomings. In particular, there is no justification for the postulates of stationary states or for the quantisation of angular momentum other than the fact that they agree with observations[2]. Furthermore, attempts to apply this model to more complicated atoms have had little success. We shall see how a quantum treatment of the atom (by solving the Schrödinger equation in 3-d) resolves these problems. Among other things, it enables us to understand atomic spectra, the periodic table, and how atoms bond to form molecules[3].

5.2. Schrödinger Equation in Spherical Coordinates

In a hydrogen atom, the orbiting electron always experiences a central force (i.e. a force directed towards the centre of the nucleus) due to Coulombic attraction between the positive nucleus and the negative electron. For this reason, it makes

[1] From this result, we see that 1d and 2d systems are not as unrealistic as they might first appear. We have seen that as the width of the square box decreases, the energy levels increase. Suppose then that a box was so tightly confined in two dimensions that the energy gap to the first excited level was more than the total energy in the system. In that case, the trapped particle could only ever exist in the ground state of those two dimensions and the dynamics in those dimensions are 'frozen out'. This effectively leaves us with a one-dimensional problem. Tricks like this are often played with trapped Bose–Einstein condensates.

[2] This is not such a bad thing really since the only justification for any postulate in science is just that it agrees with observations. However, an aim of science is to minimise the number of postulates required to explain the observable world. This is the idea of Ockham's razor, which essentially says that if we have several alternative explanations of a phenomenon, we should always plump for the simplest one.

[3] In other words, pretty much all of chemistry...

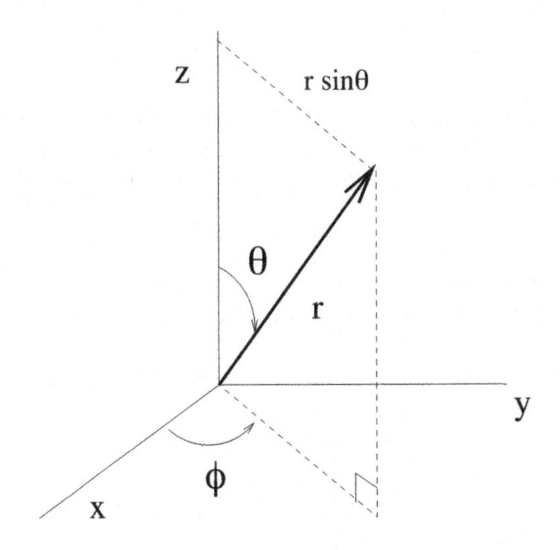

Fig. 5.1. Relations between spherical coordinates and cartesian coordinates.

the maths a lot easier if we transform to spherical coordinates, r, θ and ϕ, which are related to the cartesian coordinates x, y and z by

$$z = r\cos\theta \tag{5.5}$$

$$x = r\sin\theta\cos\phi \tag{5.6}$$

$$y = r\sin\theta\sin\phi. \tag{5.7}$$

These relations are shown in Fig. 5.1.

The transformation of the bracketed term in Eq. (5.2) to spherical coordinates is a little tedious and we will just state the result here. The interested reader can find the details in any good textbook on vector calculus. The result is

$$\frac{\partial^2\psi}{\partial x^2} + \frac{\partial^2\psi}{\partial y^2} + \frac{\partial^2\psi}{\partial z^2} = \frac{1}{r^2}\frac{\partial}{\partial r}\left(r^2\frac{\partial\psi}{\partial r}\right) + \frac{1}{r^2}\left[\frac{1}{\sin\theta}\frac{\partial}{\partial\theta}\left(\sin\theta\frac{\partial\psi}{\partial\theta}\right) + \frac{1}{\sin^2\theta}\frac{\partial^2\psi}{\partial\phi^2}\right]. \tag{5.8}$$

Substituting into Eq. (5.2) gives

$$-\frac{\hbar^2}{2Mr^2}\frac{\partial}{\partial r}\left(r^2\frac{\partial\psi}{\partial r}\right) - \frac{\hbar^2}{2Mr^2}\left[\frac{1}{\sin\theta}\frac{\partial}{\partial\theta}\left(\sin\theta\frac{\partial\psi}{\partial\theta}\right) + \frac{1}{\sin^2\theta}\frac{\partial^2\psi}{\partial\phi^2}\right] + V(r)\psi = E\psi. \tag{5.9}$$

It can also be shown (again we just state the results) that the square of the angular momentum operator of the electron, $\mathbf{L}^2$, and the z-component of the angular momentum, L_z are given by

$$\mathbf{L}^2 = -\hbar^2\left[\frac{1}{\sin\theta}\frac{\partial}{\partial\theta}\left(\sin\theta\frac{\partial}{\partial\theta}\right) + \frac{1}{\sin^2\theta}\frac{\partial^2}{\partial\phi^2}\right] \tag{5.10}$$

$$L_z = -i\hbar\frac{\partial}{\partial\phi}. \tag{5.11}$$

Despite its formidable appearance, this Schrödinger equation in spherical coordinates (5.9) can be solved in a relatively straightforward manner using the technique of separation of variables. This involves trying a solution of the form

$$\Psi(r,\theta,\phi) = \psi(r)\Theta(\theta)\Phi(\phi), \tag{5.12}$$

i.e. a product of three functions where each one is a function of only one variable. The actual working out is not important, but is given in the next section for those who are interested.

The result is that the functions $\psi(r)$, $\Theta(\theta)$ and $\Phi(\phi)$ satisfy separate eigenvalue equations. The equation for $\Phi(\phi)$ is

$$\left[-i\frac{\partial}{\partial\phi}\right]\Phi(\phi) = m\Phi(\phi). \tag{5.13}$$

The equation for $\Theta(\theta)$ is

$$\left[-\frac{1}{\sin\theta}\frac{\partial}{\partial\theta}\left(\sin\theta\frac{\partial}{\partial\theta}\right) + \frac{m^2}{\sin^2\theta}\right]\Theta(\theta) = \kappa\Theta(\theta). \tag{5.14}$$

The equation that depends on r, the so-called radial equation, is

$$\left[-\frac{\hbar^2}{2Mr^2}\frac{\partial}{\partial r}\left(r^2\frac{\partial}{\partial r}\right) + V(r) + \frac{\hbar^2\kappa}{2Mr^2}\right]\psi(r) = E\psi(r). \tag{5.15}$$

We see that these are eigenvalue equations since, in each case, an operator (in square brackets) acting on a function is equal to that same function multiplied by a constant. These three constants, m, κ and E, are the eigenvalues or quantum numbers associated with the problem. Let us now look at each of the equations (5.13–5.15) in turn.

Using (5.11), we can rewrite Eq. (5.13) as

$$L_z\Phi(\phi) = m\hbar\Phi(\phi), \tag{5.16}$$

i.e. it is just the eigenvalue equation for the z-component, $\hbar m$ of the angular momentum. We can solve (5.13) to give

$$\Phi(\phi) = \Phi(0)\exp(im\phi). \tag{5.17}$$

By requiring that the wave function does not change when $\phi \longrightarrow \phi + 2\pi$, the eigenvalues m must be integers. In other words, the z-component of the angular momentum is quantised in units of $\hbar$. This is the original Bohr hypothesis, but now it emerges quite naturally from the Schrödinger equation.

The equation for $\Theta(\theta)$ (5.14) has been studied extensively. It turns out that it has physically sensible solutions only if κ takes the form,

$$\kappa = l(l+1), \tag{5.18}$$

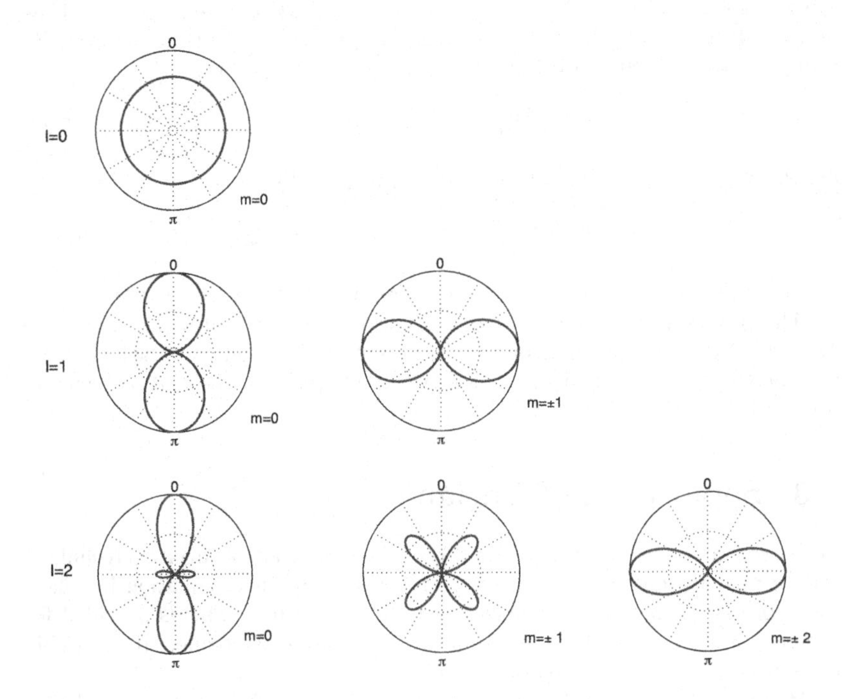

Fig. 5.2. Polar plots of the probability distributions $|Y_{lm}(\theta,\phi)|^2 = |\Theta_{lm}(\theta)|^2$. These probability distributions do not depend on ϕ.

where l is an integer $l = 0, 1, 2....$ and m is restricted to the integer values between $-l$ and $+l$. With these restrictions, the solutions of (5.14) are the associated Legendre polynomials in $\cos\theta$. Some examples are:

l	m	$\Theta_{lm}(\theta)$
0	0	1
1	0	$\cos\theta$
1	±1	$\sin\theta$
2	0	$3\cos^2\theta - 1$
2	±1	$\sin\theta\cos\theta$
2	±2	$\sin^2\theta$

(5.19)

You can check that these are solutions by direct substitution into (5.14).

These solutions may be multiplied by $\exp(im\phi)$ and still be a solution of (5.14). In fact they are now solutions of both (5.14) and (5.13) simultaneously and, apart from a constant factor, are known as the spherical harmonics. These can be looked up in many books and are generally written as, $Y_{lm}(\theta, \phi)$. They contain all the angular information about the wave function of the electron. By using equations (5.10), (5.13) and (5.14) it can be shown that,

$$\mathbf{L}^2 Y_{lm} = \hbar^2 l(l+1) Y_{lm}, \tag{5.20}$$

i.e. the angular momentum of the electron, is given by

$$L = \hbar\sqrt{l(l+1)} \qquad\qquad l = 0, 1, 2, ... \tag{5.21}$$

Proving this relationship is set as an exercise at the end of the chapter.

That leaves us with only the radial equation (5.15) to deal with. In order to solve this we need an expression for the potential $V(r)$ and so we will consider the specific case of a hydrogen atom when we pick up the thread again in Section 5.4.

5.3. Separation of Variables

In this self-contained section, we show how the separation of variables technique can be used to separate the three-dimensional Schrödinger equation in spherical coordinates into three separate eigenvalue equations. This is provided for completeness and is really intended only for the interested reader. It can safely be skipped over by everyone else.

We begin by substituting $\Psi(r, \theta, \phi) = \psi(r)\Theta(\theta)\Phi(\phi)$ into the Schrödinger equation. This gives

$$\begin{aligned} -\frac{\hbar^2}{2Mr^2}\Theta\Phi\frac{\partial}{\partial r}\left(r^2\frac{\partial\psi}{\partial r}\right) - \frac{\hbar^2}{2Mr^2}\left[\psi\Phi\frac{1}{\sin\theta}\frac{\partial}{\partial\theta}\left(\sin\theta\frac{\partial\Theta}{\partial\theta}\right)\right. \\ \left. +\psi\Theta\frac{1}{\sin^2\theta}\frac{\partial^2\Phi}{\partial\phi^2}\right] + (V(r) - E)\psi\Theta\Phi = 0. \end{aligned} \tag{5.22}$$

Dividing through by $\psi\Theta\Phi$ and multiplying by $2Mr^2$, this can be rewritten as

$$\begin{aligned} &\left[-\hbar^2\frac{1}{\psi}\frac{\partial}{\partial r}\left(r^2\frac{\partial\psi}{\partial r}\right) + 2Mr^2(V(r) - E)\right] \\ &-\hbar^2\left[\frac{1}{\Theta}\frac{1}{\sin\theta}\frac{\partial}{\partial\theta}\left(\sin\theta\frac{\partial\Theta}{\partial\theta}\right) + \frac{1}{\Phi}\frac{1}{\sin^2\theta}\frac{\partial^2\Phi}{\partial\phi^2}\right] = 0 \end{aligned} \tag{5.23}$$

Using Eq. (5.10), the second term can be rewritten to give

$$\left[-\hbar^2\frac{1}{\psi}\frac{\partial}{\partial r}\left(r^2\frac{\partial\psi}{\partial r}\right) + 2Mr^2(V(r) - E)\right] + \left[\frac{1}{\Theta\Phi}\mathbf{L}^2\Theta\Phi\right] = 0. \tag{5.24}$$

Now, the first term in square brackets depends only on r and the second term in square brackets depends only on the angular variables θ and ϕ. That means that these two terms must each equal a constant (and sum to zero), i.e. we can write

$$-\hbar^2 \frac{1}{\psi}\frac{\partial}{\partial r}\left(r^2 \frac{\partial \psi}{\partial r}\right) + 2Mr^2(V(r) - E) = -c \tag{5.25}$$

$$\frac{1}{\Theta\Phi}\mathbf{L}^2\Theta\Phi = c, \tag{5.26}$$

where c is a constant. The second equation is just the eigenvalue equation for $\mathbf{L}^2$, i.e. $c = \hbar^2 l(l+1)$, where $l = 0, 1, 2, \ldots$. This means we can rewrite the radial equation as

$$\left[-\frac{\hbar^2}{2Mr^2}\frac{\partial}{\partial r}\left(r^2\frac{\partial}{\partial r}\right) + V(r) + \frac{\hbar^2 l(l+1)}{2Mr^2}\right]\psi(r) = E\psi(r). \tag{5.27}$$

This is the same as Eq. (5.15) above. Now, using the value for c and substituting back the derivative form of $\mathbf{L}^2$ into Eq. (5.26), we get,

$$-\frac{1}{\Theta}\frac{1}{\sin\theta}\frac{\partial}{\partial\theta}\left(\sin\theta\frac{\partial\Theta}{\partial\theta}\right) - \frac{1}{\Phi}\frac{1}{\sin^2\theta}\frac{\partial^2\Phi}{\partial\phi^2} = l(l+1). \tag{5.28}$$

There is no $\Phi(\phi)$ dependence on the right hand side, i.e. $l(l+1)$, therefore, the ϕ-dependent term on the left hand side must be a constant

$$\frac{1}{\Phi}\frac{\partial^2\Phi}{\partial\phi^2} = -m^2. \tag{5.29}$$

This can be rewritten as

$$\left[-i\frac{\partial}{\partial\phi}\right]\Phi(\phi) = m\Phi(\phi), \tag{5.30}$$

i.e. Eq. (5.13).

Finally, substituting this into Eq. (5.28) and multiplying by $\Theta(\theta)$, we get

$$\left[-\frac{1}{\sin\theta}\frac{\partial}{\partial\theta}\left(\sin\theta\frac{\partial}{\partial\theta}\right) + \frac{m^2}{\sin^2\theta}\right]\Theta(\theta) = l(l+1)\Theta(\theta), \tag{5.31}$$

i.e. Eq. (5.14).

5.4. The Hydrogen Atom

So far we have solved the 3-d Schrödinger equation for the angular coordinates, θ and ϕ. These coordinates are independent of the potential $V(r)$. In order to complete the job and write down the full solution, we need to solve the radial equation

$$\left[-\frac{\hbar^2}{2Mr^2}\frac{\partial}{\partial r}\left(r^2\frac{\partial}{\partial r}\right) + V(r) + \frac{\hbar^2\kappa}{2Mr^2}\right]\psi(r) = E\psi(r). \tag{5.32}$$

For this we will consider the specific case of the electron in a hydrogen atom. In this case, the electron is bound to the proton by the Coulomb potential

$$V(r) = -\frac{e^2}{4\pi\epsilon_0 r}. \tag{5.33}$$

This enables us to rewrite the radial equation as

$$\left[-\frac{\hbar^2}{2Mr^2}\frac{\partial}{\partial r}\left(r^2\frac{\partial}{\partial r}\right) - \frac{e^2}{4\pi\epsilon_0 r} + \frac{\hbar^2 l(l+1)}{2Mr^2}\right]\psi(r) = E\psi(r), \tag{5.34}$$

where we have written $\kappa = l(l+1)$ since we have seen that these are the only values that are consistent with the solutions of the angular equations.

It turns out that this equation only has solutions if the energy eigenvalue takes the discrete values,

$$E \equiv E_n = -\frac{R}{n^2} \qquad n = 1, 2, 3, ... \tag{5.35}$$

and $l < n$. R is the Rydberg constant

$$R = \frac{1}{2}\left(\frac{e^2}{4\pi\epsilon_0}\right)^2 \frac{M}{\hbar^2} \approx 13.6\text{eV} \tag{5.36}$$

that we saw in Section 2.5. Again, we see that Bohr's energy level spectrum has emerged as a natural consequence of the Schrödinger equation.

The solutions of Eq. (5.34) for $\psi(r) \equiv \psi_{nl}(r)$ can be obtained analytically but we will just quote the results. Some un-normalised examples are:

$$\begin{array}{ccc} n & l & \psi_{nl}(r) \\ 1 & 0 & \exp\left(-\frac{r}{a_0}\right) \\ 2 & 0 & \left(1-\frac{r}{2a_0}\right)\exp\left(-\frac{r}{2a_0}\right) \\ 2 & 1 & \frac{r}{a_0}\exp\left(-\frac{r}{2a_0}\right) \end{array} \tag{5.37}$$

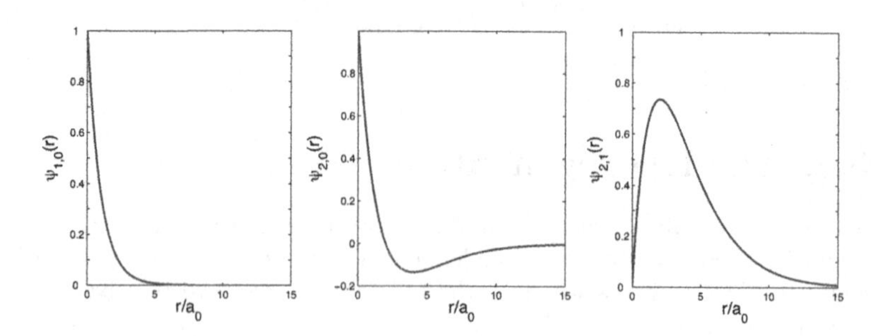

Fig. 5.3. Plot of the radial wave functions shown in (5.37).

where

$$a_0 = \frac{4\pi\epsilon_0\hbar^2}{Me^2} \approx 5.29 \times 10^{-11}\text{m} \tag{5.38}$$

is the Bohr radius. These radial wave functions are plotted in Fig. 5.3.

Let us pause to quickly recap. The complete solution of the Schrödinger equation for a Coulomb potential has the form

$$\Psi_{nlm}(r,\theta,\phi) = \psi_{nl}(r)\Theta_{lm}(\theta)\Phi_m(\phi) \tag{5.39}$$

and is labelled by three quantum numbers n, l, m. These reflect the fact that for a central potential, the original equation could be factorised into three separate equations – one for each dimension. The complete un-normalised wave functions are then easily written down by using the results in (5.19) and (5.37) and the fact that $\Phi_m(\phi) = \exp(im\phi)$. For the first two energy levels, i.e. $n = 1, 2$, the complete wave functions are:

$$\begin{array}{cccc} n & l & m & \psi_{nlm}(r,\theta,\phi) \\ 1 & 0 & 0 & \exp\left(-\frac{r}{a_0}\right) \\ 2 & 0 & 0 & \left(1-\frac{r}{2a_0}\right)\exp\left(-\frac{r}{2a_0}\right) \\ 2 & 1 & 0 & \frac{r}{a_0}\exp\left(-\frac{r}{2a_0}\right)\cos\theta \\ 2 & 1 & \pm 1 & \frac{r}{a_0}\exp\left(-\frac{r}{2a_0}\right)\sin\theta\exp(\pm i\phi). \end{array} \tag{5.40}$$

Of course, not all values of n, l, m are possible. The energy eigenvalue E_n depends only on one quantum number n also called the principal quantum number: $n = 1, 2, 3, \ldots$. The angular momentum states are labelled by quantum numbers l and m, where the total angular momentum is $\hbar\sqrt{l(l+1)}$ and the component along the z-axis is $m\hbar$. The values that l and m can take are

$$\begin{aligned} l &= 0, 1, 2, \ldots, n-1 \\ m &= -l, -l+1, \ldots, l. \end{aligned}$$

There is an alternative labelling of the angular momentum states that is commonly used and which you should know. This arose from the empirical classification of spectral lines into series before quantum theory was developed. In this notation, the angular momentum state $l = 0$ is specified by the letter s for *sharp*; $l = 1$ is specified by p for *principal*; $l = 2$ by d for *diffuse*; $l = 3$ by f for *fundamental*. After that, the letters just increase alphabetically.

$$\begin{array}{lccccccc} l = & 0 & 1 & 2 & 3 & 4 & 5 & 6 \ \ldots \\ & s & p & d & f & g & h & i \ \ldots \end{array}$$

For example, in this notation, an electron in the energy state $n = 2$ and angular momentum state $l = 0$ would be said to be in the $2s$ state. For $n = 4$ and $l = 2$, it is the $4d$ state.

5.5. Radial Probability Densities

The probability of finding an electron in a small volume dV about the point specified by the coordinates r, θ, and ϕ is proportional to:
(a) the volume dV
(b) the modulus squared of the wave function at that point, i.e. $|\psi(r,\theta,\phi)|^2$.
The volume element is spherical polar coordinates can be constructed from a small 'cube' with sides dr, $rd\theta$, and $r\sin\theta\,d\phi$, so that,

$$dV = r^2 \sin\theta\, dr\, d\theta\, d\phi. \tag{5.41}$$

Therefore, the probability of finding an electron in an energy state, n and an angular momentum state (l,m) in a volume dV at position (r,θ,ϕ) is,

$$\begin{aligned} P_{nlm}\,dV &= |\Psi_{nlm}(r,\theta,\phi)|^2 r^2 \sin\theta\, dr\, d\theta\, d\phi \\ &= \left[|\psi_{nl}(r)|^2 r^2\, dr\right]\left[|\Theta_{lm}(\theta)|^2 \sin\theta\, d\theta\right]\left[|\Phi_m(\phi)|^2\, d\phi\right], \end{aligned}$$

where in the last line we have separated the variables into each of the three sets of brackets. This means that we can normalise the wave function for the electron by evaluating three separate integrals. For example, an electron in the ground state of the hydrogen atom has the quantum numbers $n = 1$, $l = 0$, $m = 0$ and the corresponding wave function

$$\Psi_{1,0,0}(r,\theta,\phi) = A\exp\left(-\frac{r}{a_0}\right). \tag{5.42}$$

The normalisation constant A can be found as follows:

$$\begin{aligned} \int_0^\infty A^2 \exp\left(\frac{-2r}{a_0}\right) r^2\, dr \int_0^{2\pi} d\phi \int_0^\pi \sin\theta\, d\theta &= 1 \\ A^2\left(\frac{a_0^3}{4}\right)(4\pi) &= 1 \\ \Longrightarrow A &= \frac{1}{\sqrt{\pi}a_0^{3/2}}, \end{aligned} \tag{5.43}$$

where we have used the result $\int_0^\infty r^2 \exp(-r/b)\, dr = 2b^3$. This can be obtained using integration by parts and is set as an exercise at the end of the chapter.

Using a similar technique, the remaining wave functions in (5.40) can be normalised to give

$$\begin{array}{cccc} n & l & m & \Psi_{nlm}(r,\theta,\phi) \\ 1 & 0 & 0 & \frac{1}{\sqrt{\pi}a_0^{3/2}}\exp\left(-\frac{r}{a_0}\right) \\ 2 & 0 & 0 & \frac{1}{2\sqrt{2\pi}a_0^{3/2}}\left(1-\frac{r}{2a_0}\right)\exp\left(-\frac{r}{2a_0}\right) \\ 2 & 1 & 0 & \frac{1}{4\sqrt{2\pi}a_0^{3/2}}\frac{r}{a_0}\exp\left(-\frac{r}{2a_0}\right)\cos\theta \\ 2 & 1 & \pm1 & \frac{1}{8\sqrt{2\pi}a_0^{3/2}}\frac{r}{a_0}\exp\left(-\frac{r}{2a_0}\right)\sin\theta\exp(\pm i\phi). \end{array} \tag{5.44}$$

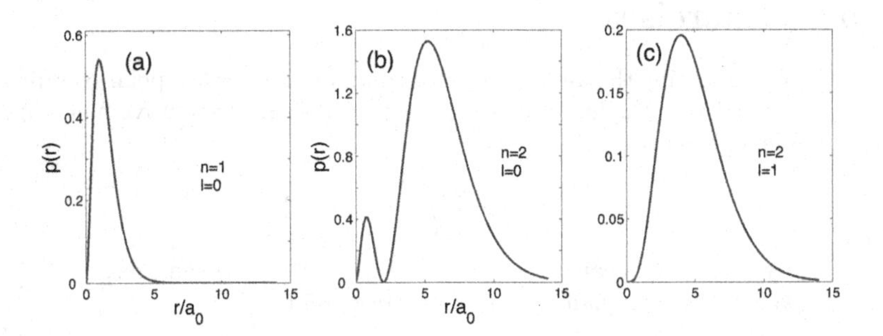

Fig. 5.4. Plot of the radial probability densities for (a) the 1s ground state, (b) the 2s state and (c) the 2p state.

If we are only interested in the distance of the electron from the nucleus and not its direction, then we can integrate over the angles, leaving the probability of finding an electron at a distance r. The radial probability density, $p(r)$, is given by

$$p(r) = |\psi_{nl}(r)|^2 r^2, \tag{5.45}$$

i.e. the modulus squared wave function is multiplied by the geometric factor r^2. For the ground state: $n = 1$, $l = 0$, $E_1 = -13.6$eV, the wave function is

$$\psi_{nl}(r) = A \exp\left(-\frac{r}{a_0}\right). \tag{5.46}$$

The normalisation constant is given by

$$A^2 \int_0^{\infty} r^2 \exp\left(-\frac{2r}{a_0}\right) dr = 1. \tag{5.47}$$

This gives $A^2 = 4/a_0^3$ and so the radial probability density is

$$p(r) = \left(\frac{4}{a_0^3}\right) r^2 \exp\left(-\frac{2r}{a_0}\right). \tag{5.48}$$

This function is plotted in Fig. 5.4(a) and peaks at the Bohr radius, $r = a_0$ – just as we would expect. For comparison, the radial distribution functions for the $n = 2$ states are plotted in Fig. 5.4(b-c).

5.6. Exercises

1. When solving the 3-d Schrödinger equation in spherical polar coordinates, we arrive at the following equation for the part of the wave function that depends on ϕ,

$$-i\frac{d\Phi(\phi)}{d\phi} = m\Phi(\phi).$$

By solving this equation, determine what values m can take.
What physical quantity does m correspond to?

2. Prove the result shown in Eq. (5.20), where $Y_{lm} = \Theta(\theta)\Phi(\phi)$.

3. Show that an electron in the $2p$ state is most likely to be found at $r = 4a_0$.

4. Evaluate the integral $\int_0^\infty r^2 \exp(-2r/a_0)\, dr$ using integration by parts.

5. Consider the wave function for the electron in a hydrogen atom. Sketch how the probability density varies with θ (i.e. $|\Theta_{lm}(\theta)|$) on a polar plot for the (i) 1s, (ii) 2p and (iii) 3d states.

6. The wave functions for the hydrogen atom can be written in terms of spherical polar coordinates (r, θ, ϕ) as

$$\Psi_{nlm}(r, \theta, \phi) = \psi_{nl}(r)\Theta_{lm}(\theta)\Phi_m(\phi).$$

(a) Identify the variables corresponding to the quantum numbers n, l and m and write down the range of values for each one. What other quantum number is needed to complete the description of the state of the electron?

(b) Normalise the following wave function for an electron in the $1s$ state

$$\psi_{100}(r, \theta, \phi) \propto \exp\left(-\frac{r}{a_0}\right).$$

You can assume the following result: $\int_0^\infty x^n \exp(-\alpha x)\, dx = n!\alpha^{-(n+1)}$.

(c) Using the normalised wave function found in part (b), calculate the expectation value of $\langle 1/r^2 \rangle$ for the $1s$ state.

(d) The $1s$ wave function shown in part (b) has its maximum value at $r = 0$, whereas the radial probability density for the $1s$ state peaks at $r = a_0$ and is zero at $r = 0$. Explain this apparent paradox.

7. The radius of the hydrogen nucleus is about 10^{-15}m. Calculate the probability that an electron in the $1s$ state is found within this radius, i.e. $r < 10^{-15}$m.

Chapter 6

Spin and Statistics

6.1. Stern–Gerlach Experiment

In 1922 one of the most important experiments in atomic physics was carried out by Otto Stern and Walter Gerlach at the University of Frankfurt. They wanted to measure the magnetic moment of atoms that comes about from the motion of the electrons. It is well-known that, if charges move in a circle, a magnetic field is created in a direction given by the right-hand rule. Think, for example, of an electric current moving in a coil. In a similar way, we have seen that charged electrons 'orbit' the nucleus of an atom with angular momentum $L = \hbar\sqrt{l(l+1)}$. This means that we would expect the atom to have an overall magnetic field associated with it. The experiment carried out by Stern and Gerlach was designed to measure this effect.

The experimental apparatus (shown in Fig. 6.1) consisted of a furnace producing a beam of silver atoms. This beam travelled through a vacuum and passed through collimators to ensure that it didn't spread out too much. The collimated beam was then passed through an *inhomogeneous magnetic field*, which was stronger near the north pole than the south pole. Finally, the atoms were collected on a screen.

The energy of a magnetic moment, $\vec{\mu}$, in a magnetic field, $\vec{B}$, is

$$E = -\vec{\mu}\cdot\vec{B} = -|\vec{\mu}||\vec{B}|\cos\theta, \tag{6.1}$$

where θ is the angle between $\vec{\mu}$. We see that the energy is minimised when $\vec{\mu}$ and $\vec{B}$ are aligned, i.e. $\theta = 0$. This is why magnetic moments try and align themselves with magnetic fields. The magnetic moment due to the angular momentum of the electron is

$$\vec{\mu} = \frac{-e\vec{L}}{2M_e}, \tag{6.2}$$

where M_e is the electron mass and there is a minus sign because electrons are negatively charged. If we take the direction of the magnetic field to be the

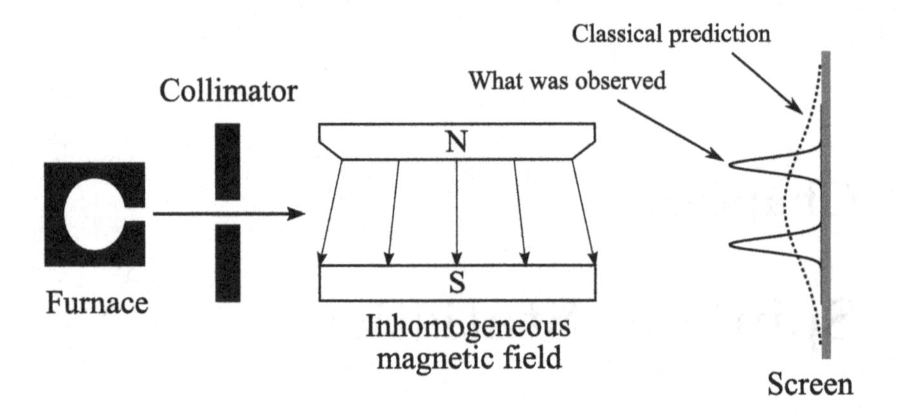

Fig. 6.1. Scheme of the Stern–Gerlach experiment. A furnace produces a source of silver atoms that are collimated and passed through an inhomogeneous magnetic field. The distribution of silver atoms is then observed on a screen. Classically we would expect to see a broad continuous distribution of atoms. In reality, two distinct peaks are observed.

z-direction, then the energy is

$$E = \frac{e}{2M_e} L_z B_z = \frac{e\hbar}{2M_e} m B_z, \tag{6.3}$$

where the last equality follows from $L_z = \hbar m$. We see that the energy depends (discretely) on the z-component of angular momentum[1]. The force on the atom in the magnetic field is then

$$F_z = -\frac{\partial E}{\partial z} = -\frac{e\hbar}{2M_e}\frac{dB_z}{dz} m. \tag{6.4}$$

Two important points come out of this. Firstly, for a uniform field, we get $dB_z/dz = 0$, which would mean that the force is zero and explains why the magnetic field needs to be inhomogeneous. This can also be physically understood by the fact that we need an overall net force, i.e. the upward and downward forces are not balanced. Secondly, we notice that the force is quantised since m can only take integer values. This second fact led Stern and Gerlach to propose that the quantisation of angular momentum would result in discrete 'blobs' of atoms being seen on the screen. In particular, there should be $2l+1$ blobs since that is the number of different values m can take. This is different from the

[1]This splitting of the energy levels in the presence of a magnetic field is called the Zeeman effect and can be used to confirm the quantisation of angular momentum. It was theoretically explained by Hendrik Lorentz, who we will meet again when we study relativity later in the book.

classical prediction of a continuous broad distribution of atoms on the screen (since classically any angular momentum is possible).

The results were not as expected — all the most interesting experiments throw up a surprise! Although they did see discrete blobs of atoms rather than a continuous distribution, they only saw two blobs. The reason this is surprising is that it would suggest $2l + 1 = 2$, which means $l = 1/2$, but we have already seen that l must be an integer. The explanation for all this is that the electron has an *inherent* magnetic moment called its spin.

6.2. What is Spin?

Spin is a vector quantity $\vec{S} = (S_x, S_y, S_z)$ measured in units of $\hbar$ and, therefore, has the units of angular momentum (as we might expect for a spin). The equation for the magnitude of the spin angular momentum is

$$S = \hbar\sqrt{s(s+1)}, \tag{6.5}$$

where s is the spin quantum number and can only be an integer or half-integer.

The word spin is somewhat misleading as nothing is spinning. For example, the electron is a point particle and a point cannot spin. Despite this, it is possible to see evidence of the spin (or inherent magnetic moment) of a particle in the Stern–Gerlach experiment. In analogy with the magnetic moment due to its orbital angular momentum, the energy of the particle in a magnetic field $\vec{B}$ due to its spin is

$$E = -g\vec{S} \cdot \vec{B}, \tag{6.6}$$

where g is a constant. If we define the z axis to be in the direction of the magnetic field, we get $\Delta E = -gS_zB$. The z-component of the spin can only take the values

$$S_z = m_s\hbar, \qquad \text{where} \quad m_s = -s, -s+1, -s+2, ..., s \tag{6.7}$$

and so the energy level is split into $2s+1$ components. This means that the spin of a particle can be inferred by its behaviour in a magnetic field. For example, a particle split into three energy components by a magnetic field would have spin 1.

At this point, the astute reader may be wondering what was wrong with the original reasoning of Stern and Gerlach. Why doesn't their experiment show evidence of the orbital angular momentum of the atom? It turns out that there is nothing wrong at all. In fact, strictly speaking, the line-splitting that is observed is due to the total angular momentum, i.e. the sum of the spin and orbital angular momenta. The total angular momentum of an atom is obtained by adding up the angular momenta of all the particles it contains[2]. It is usually

[2] It is typically a good approximation to consider only the angular momenta of the electrons, since the angular momentum of the nucleus is a small correction.

denoted by the vector $\vec{J} = \vec{L} + \vec{S}$ and its magnitude, J, obeys the quantisation rule

$$J = \hbar\sqrt{j(j+1)}, \tag{6.8}$$

where the quantum number j can take only integer or half-integer values. The z-component of the total angular momentum can only take the values

$$J_z = m_j\hbar, \qquad \text{where} \quad m_j = -j, -j+1, -j+2, ..., j. \tag{6.9}$$

Thus a Stern–Gerlach experiment using atoms with angular momentum of magnitude $J = \hbar\sqrt{j(j+1)}$ should exhibit $(2j+1)$ spots on the detection screen. So why did Stern and Gerlach only see the spin part? Well, as fate would have it, they happened to pick silver atoms for their experiment, for which the orbital angular momentum of the atom is zero. In this case, $j = s = +1/2$, so two spots were seen. Subsequent experiments with atoms of different elements have shown the quantisation of the z-component of angular momentum as well as spin. So the Stern–Gerlach experiment was well-conceived – it just turned out to be better than they (or anyone else at the time) could have imagined[3].

6.3. Symmetry of the Wave Function

So far, we have studied only the wave function for single particles. To study real-world problems, we would like to extend this approach to multi-particle systems. This leads us directly to an important feature of quantum mechanics: *quantum particles of the same type are identical.* Two particles are said to be identical when they cannot de distinguished by any intrinsic property. This means that two electrons, for example, cannot be told apart no matter how clever we are or how hard we try. This is quite different from classical physics where objects may appear to be the same (identical twins, two red apples....) but can always be distinguished if we look hard enough. While classical particles have sharp trajectories and so can be told apart by the paths they take, in quantum physics there is no way of keeping track of the individual particles when their wave functions overlap. The fact that quantum particles are indistinguishable has some fascinating and fundamental consequences. We have already seen, for example, that the phenomenon of Bose–Einstein condensation occurs when the de Broglie wavelengths of certain particles overlap and the particles can no longer be distinguished.

[3]Scientific discoveries often involve a dose of good luck and the Stern–Gerlach experiment was no exception. When Gerlach first removed the detector plate from the vacuum after this experiment was carried out, nothing was seen. There was no trace of silver. However, to his amazement, slowly the trace of the beam started to appear. As Stern later put it:

> "My salary was too low to afford good cigars, so I smoked bad cigars. These had a lot of sulphur in them, so my breath on the plate turned the silver into silver sulphide, which is jet black so easily visible. It was like developing a photographic plate."

Had it not been for Stern's cheap cigars, one of quantum physics greatest episodes may have passed us by.

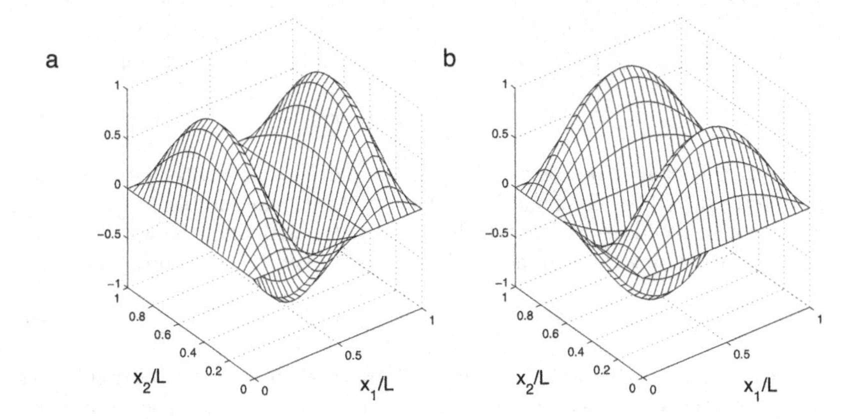

Fig. 6.2. Plot of (a) $\sin(3\pi x_1/L)\sin(\pi x_2/L)$ and (b) $\sin(3\pi x_2/L)\sin(\pi x_1/L)$.

6.4. Wave Function for Two Identical Particles

As a simple example, let us think about two identical non-interacting particles of mass M placed in an infinite square well with width L. We know that for single particles, the energy of the pth level is given by

$$E_p = \frac{\hbar^2}{2M}\left(\frac{\pi p}{L}\right)^2. \tag{6.10}$$

The spatial wave function of a particle in the pth energy level is

$$\psi(x) = \sqrt{\frac{2}{L}}\sin\left(\frac{p\pi x}{L}\right). \tag{6.11}$$

Now, suppose that one particle goes into level p and the other goes into level q. Since the particles are not interacting, they should be independent of one another and, for the wave function of the two particles, we might anticipate the product

$$\psi(x_1, x_2) = \psi_p(x_1)\psi_q(x_2) = \frac{2}{L}\sin\left(\frac{p\pi x_1}{L}\right)\sin\left(\frac{q\pi x_2}{L}\right), \tag{6.12}$$

where the particle in state p has been labelled by x_1 and the one in q labelled x_2.

It is interesting to ask what happens if the labels are interchanged $1 \leftrightarrow 2$. In this case, the new wave function becomes

$$\psi(x_2, x_1) = \psi_p(x_2)\psi_q(x_1) = \frac{2}{L}\sin\left(\frac{p\pi x_2}{L}\right)\sin\left(\frac{q\pi x_1}{L}\right), \tag{6.13}$$

and comparing with (6.12) we see that the wave function has changed, i.e. $\psi(x_1, x_2) \neq \psi(x_2, x_1)$. An example is shown in Fig. 6.2 for $p = 1$ and $q = 3$.

More importantly, the probability distribution function has changed

$$|\psi(x_1, x_2)|^2 \neq |\psi(x_2, x_1)|^2. \tag{6.14}$$

However, this cannot be correct since, if the particles are truly identical, then the interchange $1 \leftrightarrow 2$ should not affect anything measurable otherwise the measurement could be used to tell them apart. In particular, the wave function $\psi(x_1, x_2)$ describing the two identical particles should have the property that

$$|\psi(x_1, x_2)|^2 = |\psi(x_2, x_1)|^2. \tag{6.15}$$

To understand this condition let us introduce an operator $\mathcal{P}$ that swaps the two particles, i.e. $\mathcal{P}\psi(x_1, x_2) = \psi(x_2, x_1)$. Clearly, now, if we apply this operator twice, we should get the original state back (since we swap the two particles then swap them back again). This gives

$$\mathcal{P}^2\psi(x_1, x_2) = \psi(x_1, x_2), \tag{6.16}$$

and so $\mathcal{P}^2$ is just the identity and the eigenvalues of $\mathcal{P}$ are just ± 1. In other words, we have two possibilities:

$$\psi(x_1, x_2) = \psi(x_2, x_1) \tag{6.17}$$

$$\psi(x_1, x_2) = -\psi(x_2, x_1). \tag{6.18}$$

For the + sign, the wave function is said to be *symmetric* and for the − sign, it is said to be *antisymmetric*.

So it seems that the fact that quantum particles are identical, demands that the wave function is either symmetric or antisymmetric. We can easily construct symmetric and antisymmetric combinations for the two particles in a box. These are respectively:

$$\psi_S(x_1, x_2) = \frac{1}{\sqrt{2}}\left[\psi_p(x_1)\psi_q(x_2) + \psi_p(x_2)\psi_q(x_1)\right] \tag{6.19}$$

$$\psi_A(x_1, x_2) = \frac{1}{\sqrt{2}}\left[\psi_p(x_1)\psi_q(x_2) - \psi_p(x_2)\psi_q(x_1)\right] \tag{6.20}$$

The probability distribution functions derived from the correctly symmetrised and antisymmetrised wave functions are plotted in Fig. 6.3 and are quite different from those in Fig. 6.2.

Although the two particles are non-interacting, the probability distribution function for the symmetric case favours particles being close together ($x_1 \approx x_2$). Whereas the antisymmetric function keeps the particles apart since the wave function vanishes at $x_1 = x_2$. This intriguing result is a purely quantum mechanical effect and has profound consequences for the physical world.

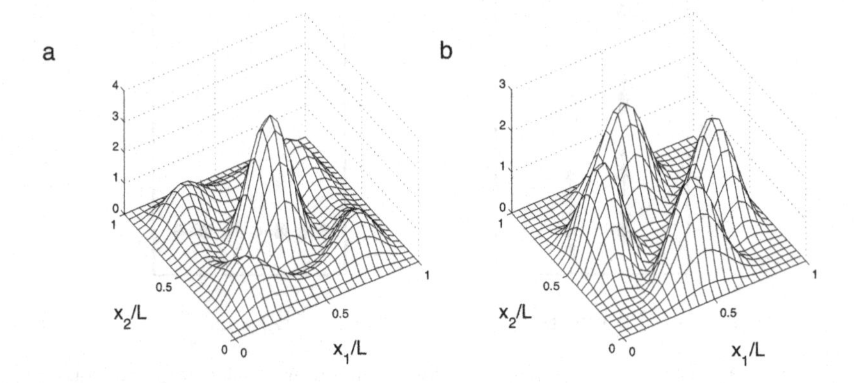

Fig. 6.3. Plot of (a) $|\psi_S(x_1, x_2)|^2$ and (b) $|\psi_A(x_1, x_2)|^2$ for $p = 1$ and $q = 3$.

6.5. The Pauli Exclusion Principle

We have seen that the wave functions of identical particles must be either symmetric or antisymmetric. The antisymmetric form has the property that if two particles are put in the same quantum state, e.g. $p = q$, the wave function is

$$\psi(x_1, x_2) = \frac{1}{\sqrt{2}} [\psi_p(x_1)\psi_p(x_2) - \psi_p(x_2)\psi_p(x_1)] = 0. \tag{6.21}$$

This means that an antisymmetric wave function prevents two particles from having the same quantum numbers. For the symmetric wave function there is no such restriction.

- Particles described by symmetric wave functions are *bosons* – named after S.N. Bose.
- Particles described by antisymmetric wave functions are *fermions* – named after E. Fermi[4].

All particles are either bosons or fermions – this is an innate characteristic of the particle and is closely related to their spin. It has been shown that particles with:

integer spin, e.g. 0, 1, 2, ... are bosons

half-integer spin, e.g. 1/2, 3/2, ... are fermions

[4]Fermi was a brilliant physicist who distinguished himself at a young age. His seeming infallibility and supreme authority on all matters of physics earned him the nickname ‘The Pope’ among his colleagues.

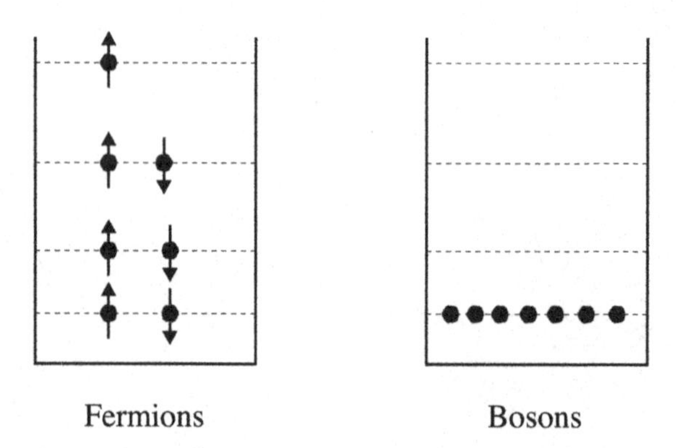

Fig. 6.4. A maximum of two spin-1/2 fermions can occupy each energy level and they must have opposite spins. There is no restriction on how many bosons can occupy a single energy level.

Systems consisting of fermions obey Fermi–Dirac statistics and systems consisting of bosons obey Bose–Einstein statistics. Electrons have spin $+\hbar/2$ (spin up) or $-\hbar/2$ (spin down) and so are fermions. Consequently, no two electrons can have the same set of quantum numbers. This is an expression of the Pauli Exclusion Principle.

Pauli Exclusion Principle: No two identical fermions can have the same set of quantum numbers.

We can understand superconductors in terms of bosons and fermions. In a superconductor, pairs of electrons are persuaded to form bound pairs (called Cooper pairs), which have an overall integer spin. These pairs then act as bosons and superconductivity appears when these pairs are all in the same quantum state. Similarly, photons are bosons and so can all occupy a single quantum state – this is what laser light is.

6.6. Spin States and Spin Functions

The two spin states of an electron, $s_z = +1/2$ and $s_z = -1/2$ are referred to as spin up and spin down respectively. The wave functions for these two spin states can be written as $\chi(\uparrow)$ and $\chi(\downarrow)$. Since the spin of a particle has nothing to do with its position in space, i.e. they are independent, the total wave function can be written as the product of the spatial and spin wave functions. For example, an electron (labelled 1) in the pth energy level with spin up can be written as

$$\psi_p(x_1)\chi(\uparrow). \tag{6.22}$$

The Pauli Exclusion Principle now allows a spin up and a spin down electron in each energy level since the extra spin quantum number enables a total antisymmetric wave function to be formed even though the spatial part is symmetric. There are two possibilities for constructing an overall wave function that is antisymmetric:

$$\psi(x_1, x_2) = (\text{symmetric spatial part}) \times (\text{antisymmetric spin part})$$
$$\psi(x_1, x_2) = (\text{antisymmetric spatial part}) \times (\text{symmetric spin part}).$$

We have seen that if two electrons are in the same energy state, then only the symmetric spatial part is possible (the antisymmetric part is zero everywhere) and hence must be in an antisymmetric spin state

$$\chi_1(\uparrow)\chi_2(\downarrow) - \chi_1(\downarrow)\chi_2(\uparrow). \tag{6.23}$$

This spin state has total spin $S = s_1 + s_2 = 0$ and is known as a singlet state. The total wave function of two particles in the same energy state is then

$$\psi(x_1, x_2) = [\psi_p(x_1)\psi_p(x_2)]\,(\chi_1(\uparrow)\chi_2(\downarrow) - \chi_1(\downarrow)\chi_2(\uparrow)). \tag{6.24}$$

and we see that the two spins must point in opposite directions.

When the electrons are in different energy levels, more possibilities are open to the form of the wave function. The combination (symmetric spatial part)×(antisymmetric spin part) gives

$$\psi(x_1, x_2) = [\psi_p(x_1)\psi_q(x_2) + \psi_p(x_2)\psi_q(x_1)]\,(\chi_1(\uparrow)\chi_2(\downarrow) - \chi_1(\downarrow)\chi_2(\uparrow)). \tag{6.25}$$

The other combination (antisymmetric spatial part)×(symmetric spin part) gives three possibilities,

$$\psi(x_1, x_2) = [\psi_p(x_1)\psi_q(x_2) - \psi_p(x_2)\psi_q(x_1)] \times \begin{cases} (\chi_1(\uparrow)\chi_2(\uparrow)) \\ (\chi_1(\uparrow)\chi_2(\downarrow) + \chi_1(\downarrow)\chi_2(\uparrow)) \\ (\chi_1(\downarrow)\chi_2(\downarrow)). \end{cases} \tag{6.26}$$

The three symmetric spin states describe a total spin state of $S = 1$. This is known as a triplet state. The existence of these spin states plays an important role in the fine structure of atomic spectra.

6.7. Bose–Einstein and Fermi–Dirac Distributions

We can write down the distribution of particles at each energy for bosons and fermions. The Bose–Einstein distribution, which applies to bosons is given by

$$n(E) = \frac{1}{e^{(E-\mu)/kT} - 1}, \tag{6.27}$$

for $E > \mu$ and where $n(E)$ is the number of particles in a state with energy E, k is Boltzmann's constant, T is the temperature and μ is the chemical potential. The chemical potential can be thought of as the energy required to add one particle to the distribution[5] and, in practice, it can be thought of as a constant that can be varied to fix the total number of particles in the system, i.e.

$$N = \int_0^\infty g(E)n(E)\,dE, \tag{6.28}$$

where $g(E)$ is the density of states. In the particular case of photons, the number of particles is not conserved since photons can be readily created and destroyed. In this case, the chemical potential is zero and the Bose–Einstein distribution can be written as

$$n(f) = \frac{1}{e^{hf/kT} - 1}, \tag{6.29}$$

where we have used $E = hf$ for a photon of frequency, f. The energy at each frequency is then given by,

$$E(f) = hfn(f) = \frac{hf}{e^{hf/kT} - 1}, \tag{6.30}$$

which is precisely the expression that Planck guessed (2.2) when trying to explain black body radiation.

By contrast, the Fermi–Dirac distribution, which describes fermions is

$$n(E) = \frac{1}{e^{(E-\mu)/kT} + 1}. \tag{6.31}$$

In this case, the chemical potential is often called the Fermi energy, E_F. Physically, it can be thought of as the energy of the most energetic particle out of a collection of fermions at zero temperature. Making this substitution, the Fermi–Dirac distribution can be written as

$$n(E) = \frac{1}{e^{(E-E_F)/kT} + 1}. \tag{6.32}$$

From this we see that, whatever the temperature, $n(E_F)$ is always equal to 1/2. We also see that $n(E)$ is always less than or equal to 1. This is consistent with the Pauli exclusion principle, which states that we can have at most one fermion in each state. By contrast, for the Bose–Einstein distribution (6.27), there is no limit to the number of particles at each energy. In particular, as $E \to \mu^+$, $n(E) \to \infty$.

In Fig. 6.5 we compare the Bose–Einstein and Fermi–Dirac distributions with the classical Maxwell-Boltzmann distribution, $n(E) = \exp(-(E - \mu)/kT)$ at temperature $T = 5000\text{K}$ and chemical potential $\mu = 2\text{eV}$. At large energies, all three distributions agree since the quantum effects become unimportant and the gas behaves classically in this limit.

We can now use what we know about Fermi–Dirac statistics to understand some of the properties of electrons in a metal. Metals are excellent electrical

[5] Strictly it is the energy required to add one particle to the distribution at constant volume and entropy.

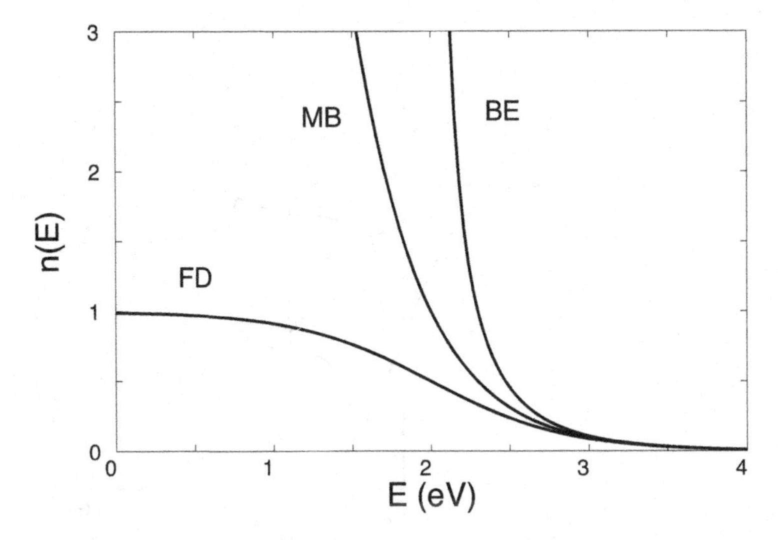

Fig. 6.5. Comparison of the Bose–Einstein (BE), Fermi–Dirac (FD) and Maxwell-Boltzmann (MB) distributions at temperature $T = 5000\text{K}$ and chemical potential $\mu = 2\text{eV}$.

conductors because the valence electrons are only weakly bound to atoms. This means that, to a good approximation, we can treat the valence electrons simply as a gas of electrons trapped in a container given by the surface of the metal. Such an approach is called the free electron model of metals and was developed by Arnold Sommerfeld. Despite its simplicity, this model has been remarkably successful in explaining many experimental phenomena.

Let us start by thinking about the behaviour of the Fermi–Dirac distribution given by (6.31). This distribution is shown in Fig. 6.6 for $T = 0$ and $T > 0$. For $T = 0$, we have $n(E < E_F) = 1$ and $n(E > E_F) = 0$. In other words, each state is occupied up to the Fermi energy and, beyond that, each state is empty. A conduction electron in the metal with the energy E_F has a velocity[6]

$$v_F = \sqrt{2E_F/M_e}, \tag{6.33}$$

where M_e is the mass of the electron and v_F is called the Fermi velocity. Oddly, this means that for a metal with a Fermi energy of about 4eV (which is a typical value), the electrons at the Fermi surface have a velocity of over 1000 km/s even at zero temperature. Pauli's exclusion principle simply doesn't allow them to have smaller velocities.

What happens as the temperature is increased? In that case, electrons can

[6]It is perfectly reasonable to treat the electrons as nonrelativistic and use Newtonian mechanics.

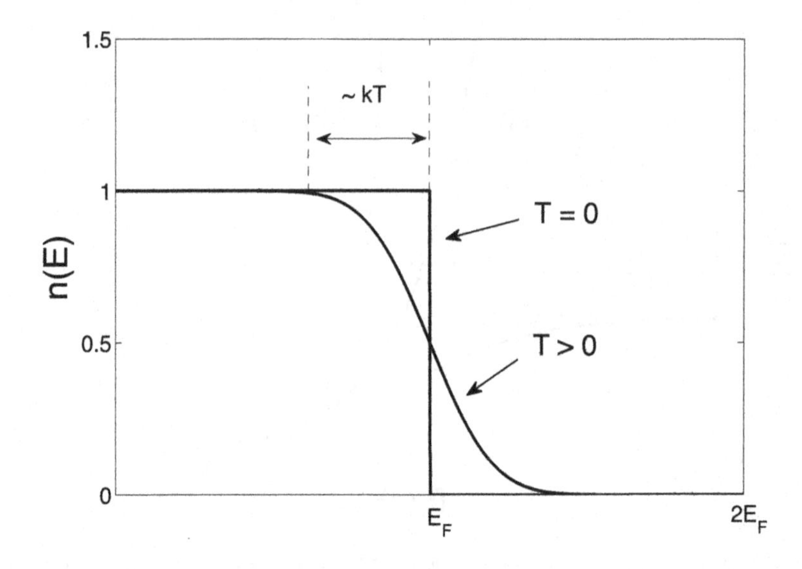

Fig. 6.6. The Fermi–Dirac distribution at $T = 0$ and $T > 0$.

gain energy of about kT. This means that electrons that are already within kT of the Fermi energy can be excited to energy levels greater than the Fermi energy. Electrons with energies much less than the Fermi energy, i.e. $E \ll E_F$ will not be excited because they cannot gain enough thermal energy to reach the next available unoccupied state. This means that for $0 < kT << E_F$, the Fermi–Dirac distribution gets 'smoothed off' near the Fermi energy as shown in Fig. 6.6.

Let us now calculate the Fermi energy for real metals using the free electron model. The density of states can be shown to be

$$g(E)dE = \frac{\sqrt{2}M_e^{3/2}}{\hbar^3\pi^2}\sqrt{E}\,dE. \tag{6.34}$$

From this we can write the electron density, N/V, where N is the total number of free electrons and V is the volume of the metal, as

$$N/V = \int_0^\infty g(E)n(E)\,dE = \frac{\sqrt{2}M_e^{3/2}}{\hbar^3\pi^2}\int_0^\infty \frac{\sqrt{E}}{e^{(E-E_F)/kT}+1}\,dE. \tag{6.35}$$

This integral is very easy to evaluate for the case $T = 0$. This is because the denominator of the integrand, in that case, is simply unity for $E < E_F$ and infinite for $E > E_F$ as we saw in Fig. 6.6. This gives us

$$N/V = \frac{\sqrt{2}M_e^{3/2}}{\hbar^3\pi^2}\int_0^{E_F} \sqrt{E}\,dE$$

$$= \frac{2\sqrt{2}M_e^{3/2}}{3\hbar^3\pi^2} E_F^{3/2}. \tag{6.36}$$

Rearranging, we can find an expression for the Fermi energy at zero temperature[7]

$$E_F = \frac{\hbar^2}{2M_e}\left(\frac{3\pi^2 N}{V}\right)^{2/3}. \tag{6.37}$$

From this result we see that the Fermi energy increases with the electron density N/V. This is what we would expect since Pauli's exclusion principle tells us that we can only have two electrons (with opposite spin) in each level. So, if we were to put more and more electrons into a metal with fixed volume, the electrons would be forced to occupy ever higher energy levels, thereby increasing the Fermi energy.

Finally, it is worth commenting that the free electron model explains why electrons seem to contribute an anomalously small amount to the heat capacity of a metal. If the electrons behaved classically then heating a metal from absolute zero to a temperature T should result in the average energy of each electron increasing by $3kT/2$. However, since the electrons obey a Fermi–Dirac distribution, we have seen (e.g. in Fig. 6.6) that only the electrons near the Fermi energy can be excited. This means that a much smaller amount of energy is absorbed by the electrons than would be expected from a classical model. This is a result that can only be explained by quantum statistics.

[7]The Fermi energy depends on temperature but only weakly. It is a reasonable approximation to say that the Fermi energy at temperature T is equal to the Fermi energy at zero temperature for T up to a few thousand K.

6.8. Exercises

1. A one dimensional box of width 1nm contains 11 electrons. The system is in the ground state.

   (a) What is the Fermi energy (i.e. energy of the most energetic fermion) assuming the electrons are non-interacting?

   (b) What is the minimum energy, in eV, a photon must have to excite an $n = 1$ electron to a higher energy state?

2. Consider two identical non-interacting spin-1/2 fermions. Write down the possible total wave functions (including both the spatial and the spin components) and group these wave functions into a singlet and a triplet.

3. Use the free electron model to show that the average energy of a conduction electron in a metal at temperature $T = 0$ is given by $E = 3E_F/5$.

4. Copper has a molar mass of 63.54 g/mol and a density of 8.94 g/cm^3. Given that each copper atom provides one conduction electron, calculate:
   (a) the density of conduction electrons
   (b) the Fermi energy
   (c) the Fermi velocity.

Chapter 7

Atoms, Molecules and Lasers

7.1. Periodic Table

Now that we have a full quantum description of electrons in an atom as well as an understanding of how multi-electron systems behave, we should be able to study the structure of atoms that are more complicated than hydrogen. In principle, we can do this simply by finding the wave function for each electron separately. Of course, in reality there will be interactions between the different electrons that will change things a bit. However, it turns out to be a good approximation to ignore the interactions, and that is the approach we will take here.

The state of an electron in an atom is fully specified by the four quantum numbers: n, l, m and m_s[1]. This, combined with the fact that we know that electrons are fermions and so must obey the Pauli exclusion principle, allows us to study the electron configurations of different atoms. In particular, it enables us to understand the periodic table of the elements.

In 1867, the Russian chemist Dmitri Mendeléev was the first to propose a periodic arrangement of the elements. He did so by explicitly pointing out 'gaps' where undiscovered elements should exist. He could then predict the expected properties of the missing elements. The subsequent discovery of these elements verified Mendeléev's scheme and element number 101 has since been named in his honour[2].

A version of the periodic table is shown in Fig. 7.1. The periodic table is a useful tool for predicting the behaviour of different elements. However, its

[1]Strictly there is a fifth quantum number, s, which is the spin of the particle. However, since this always has a value of $s = 1/2$ for an electron, there is no need to explicitly state it.

[2]Intriguingly, the periodic table supposedly came to Mendeléev in a dream. Upon waking, he quickly wrote down the elements as they had arranged themselves in his dream and, amazingly, only one correction had to be made later.

1 H																	2 He
3 Li	4 Be											5 B	6 C	7 N	8 O	9 F	10 Ne
11 Na	12 Mg											13 Al	14 Si	15 P	16 S	17 Cl	18 Ar
19 K	20 Ca	21 Sc	22 Ti	23 V	24 Cr	25 Mn	26 Fe	27 Co	28 Ni	29 Cu	30 Zn	31 Ga	32 Ge	33 As	34 Se	35 Br	36 Kr
37 Rb	38 Sr	39 Y	40 Zr	41 Nb	42 Mo	43 Tc	44 Ru	45 Rh	46 Pd	47 Ag	48 Cd	49 In	50 Sn	51 Sb	52 Te	53 I	54 Xe
55 Cs	56 Ba	57 La	72 Hf	73 Ta	74 W	75 Re	76 Os	77 Ir	78 Pt	79 Au	80 Hg	81 Tl	82 Pb	83 Bi	84 Po	85 At	86 Rn
87 Fr	88 Ra	89 Ac	104 Rf	105 Db	106 Sg	107 Bh	108 Hs	109 Mt	110 Ds	111 Rg	112 Cn	113 Uut	114 Uuq	115 Uup	116 Uuh		

58 Ce	59 Pr	60 Nd	61 Pm	62 Sm	63 Eu	64 Gd	65 Td	66 Dy	67 Ho	68 Er	69 Tm	70 Yb	71 Lu
90 Th	91 Pa	92 U	93 Np	94 Pu	95 Am	96 Cm	97 Bk	98 Cf	99 Es	100 Fm	101 Md	102 No	103 Lr

Fig. 7.1. Periodic table of the elements.

significance and beauty for physicists lies in the fact that it demonstrates that there is a regularity to the *structure* of atoms. Quantum mechanics successfully explains this structure. We need three basic ideas to see how this works:

1. The energy levels of the atom are found by solving the Schrödinger equation for multielectron atoms and depend on the quantum numbers n and l.

2. For each value of l, there are $2l + 1$ possible values for m and for each of these there are two possible values for the spin quantum number, m_s. Consequently, each value of l has $2(2l + 1)$ different states associated with it and each of these have the same energy. This can be summarised as

Subshell	l	**Number of states**
s	0	2
p	1	6
d	2	10
f	3	14

(7.1)

3. The ground state of the atom is the lowest-energy electron configuration that is consistent with the Pauli exclusion principle.

The energy of the electron is determined mainly by the principal quantum number n (which we have seen is related to the radial dependence of the wave function) and by the orbital angular-momentum quantum number, l. Generally, the lower the value of n, the lower the energy and, for a given value of n, the lower the value of l, the lower the energy (though there are exceptions). The dependence of the energy on l is due to the interaction of the electrons in the atom with each other. In hydrogen, of course, there is only one electron, and the energy is independent of l.

The electron structure of an atom is written as a listing of all the energy levels that the electrons are in, ordered by their energy. For example, the ground state of hydrogen is $1s^1$ denoting one electron in the $1s$ state. The ground state for helium is $1s^2$ and for lithium is $1s^2 2s^1$. Remember that the s state can only be occupied by two electrons, because there are only two spin states, and so the third electron in lithium must go to the next lowest available level – in this case $2s$. The energy ordering of the levels is as follows,

$$1s < 2s < 2p < 3s < 3p < 4s < 3d < 4p. \tag{7.2}$$

Note that the order of $4s$ and $3d$ are different from what we might guess.

Now we can write down the electron configurations for much more complicated atoms. For example, the ground state configuration for arsenic (As) with atomic number 33 is,

$$1s^2 2s^2 2p^6 3s^2 3p^6 4s^2 3d^{10} 4p^3.$$

If the electron configuration for an atom is

$$1s^2 2s^2 2p^6 3s^2 3p^6 4s^2 3d^9 4p^4,$$

the atom is still arsenic (33 electrons). However, this is now an excited state of arsenic since one of the $3d$ electrons has moved to the more energetic $4p$ level.

7.2. Ionisation Energies

The ionisation energy is the energy required to remove a ground-state electron from an atom and leave a positive ion behind. The ionisation energy of hydrogen is 13.6 eV because the ground-state energy is $E_0 = -13.6\,\text{eV}$ and an electron requires a minimum of zero energy to be unbound. A plot of the ionisation energy for different elements is shown in Fig. 7.2 and we see there is a clear pattern to the values. The ionisation energies for the alkali metals (e.g. Li, Na, K) on the left of the periodic table have low values. These increase steadily across the table to reach maxima for the inert gases (e.g. He, Ne, Ar) on the right hand side of the table.

Quantum theory can explain this structure. The inert gases have closed shells, which are very stable structures and so these elements are chemically non-reactive. It takes a lot of energy to pull an electron out of a stable closed shell and so the inert gases have the largest ionisation energies. The alkali metals, by contrast, have a single s-electron outside a closed shell. This electron is easily stripped away, which is why these elements are highly reactive and have the lowest ionisation energies.

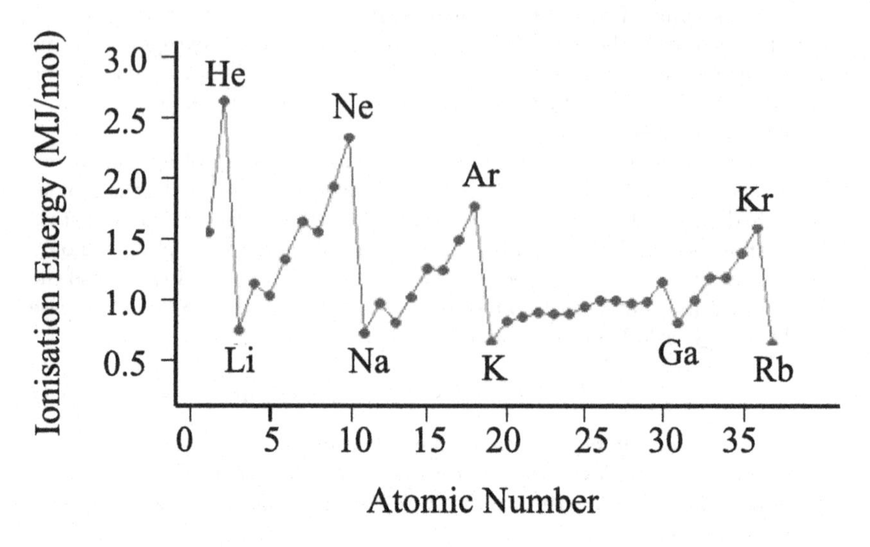

Fig. 7.2. Ionisation energy of the elements.

We can also understand some of the other features of the ionisation energy graph. As we move from left to right across the periodic table, we see that there is a drop in the ionisation energy between the second and third elements. For example after Li ($Z = 3$), the ionisation energy increases for Be ($Z = 4$), but then decreases for B ($Z = 5$). The reason for this is that the Be has a full s-orbital and so the next electron for B must go into a p-orbital. Since the p orbital has a higher energy, it is easier to strip this electron away from the atom, hence the ionisation energy decreases.

After the next three elements, there is another dip in the ionisation energy – the same is true as we move from Na to Ar. This effect can be explained by the interaction between the electrons. We know that there can be six electrons in the p-orbital: there are three levels corresponding to the different values that m can take, i.e. $m = -1, 0, +1$, and each level can be occupied by two electrons with opposite spins. As we saw when discussing the Stern–Gerlach experiment, electrons prefer to have their spins aligned since this is the lowest energy configuration. What this means is that, as we start filling the p-orbital, the first three electrons have their spins aligned and, therefore, go into levels corresponding to different values of m. The fourth electron is then forced to go into a level that is already occupied and so must have the opposite spin. This means that it has a higher energy and so the ionisation energy decreases. This effect is known as Hund's rule.

7.3. Energy Spectrum

Left to itself, an atom will typically be found in its lowest-energy (i.e. ground) state. An atom can get into an excited state by absorbing energy. This can happen, for example, by absorbing a photon or by colliding with other atoms.

One of the postulates of Bohr's model is that an atom can jump from one stationary state of energy E_0 to a higher energy state E_1 by absorbing a photon of frequency,

$$f = \frac{\Delta E_{\text{atom}}}{h} = \frac{E_1 - E_0}{h}. \tag{7.3}$$

Because we are interested in spectra, it is often more useful to write this equation in terms of wavelength. Since eV is the natural energy scale and nm is the natural wavelength scale for atomic spectra, a convenient form of the formula is

$$\lambda = \frac{c}{f} = \frac{hc}{\Delta E_{\text{atom}}} = \frac{1240 \text{ eV nm}}{\Delta E(\text{in eV})}. \tag{7.4}$$

Bohr's idea of quantum jumps remains an integral part of our interpretation of the results of quantum mechanics. By absorbing a photon, an atom jumps from its ground state to one of its excited states. However, it turns out that not every conceivable transition can occur – some are 'forbidden' (i.e. highly improbable). Allowed transitions between energy states, with the emission or absorption of a photon, are governed by the following selection rules,

$$\Delta l = \pm 1$$
$$\Delta m = 0, \pm 1.$$

These are the selection rules for an electric dipole transition[3]. For example, an atom in an s-state ($l = 0$) can absorb a photon and be excited to a p-state ($l = 1$), but not to another s-state or to a d-state. These rules come naturally from the conservation of angular momentum and the fact that a photon carries one unit, $\hbar$, of angular momentum since it is a spin-1 particle. This enables us to understand the spectra of atoms.

7.4. Ionic Bonding

The simplest type of bond is the ionic bond found in salts, such as sodium chloride (NaCl) and is due to the electrostatic attraction of two oppositely charged ions. Sodium, for example, has a solitary 3s electron located outside a stable electron core. This electron can be stripped away from the atom with the relatively small energy of 5.14 eV. Chlorine by contrast is one electron short of a full shell and *releases* 3.62 eV when it acquires an additional electron. Thus the formation of a Na^+ ion and a Cl^- ion by the donation of one electron from Na to Cl requires only 5.14 eV - 3.62 eV = 1.52 eV at infinite separation. Although this is only a modest amount of energy, you may wonder why ionic

[3]Note that electron spin-flips are forbidden in such a transition, i.e. $\Delta m_s = 0$.

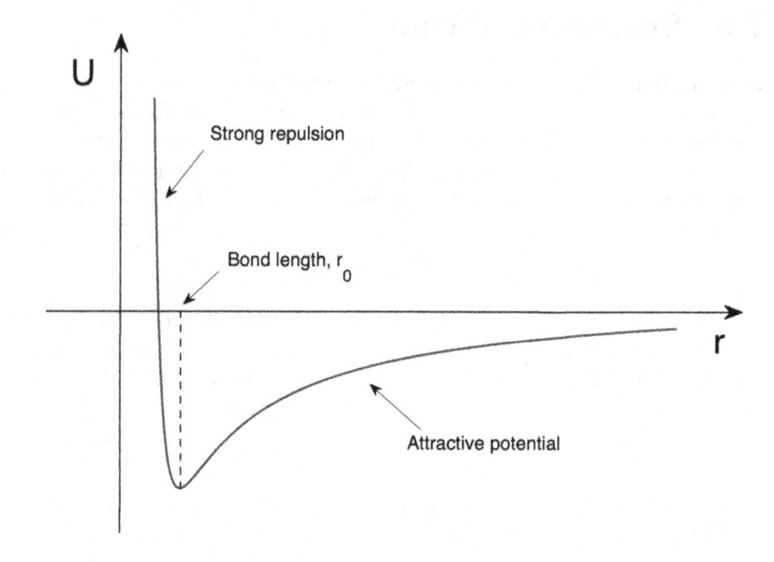

Fig. 7.3. Variation of the potential energy as a function of the separation between the ions in an ionic bond.

bonds form at all since there is still apparently an energy cost associated with them. However, we have not yet accounted for the energy 'payback' that we get from the attraction of the resulting ions.

The electrostatic energy of the oppositely charged ions when they are separated by a distance r is $-e^2/(4\pi\epsilon_0 r)$. When the separation of the ions is less than about 0.95 nm, the negative potential energy of attraction is greater than the 1.52 eV needed to create the ions. It, therefore, becomes energetically favourable for the two atoms to exchange an electron and form an ionic bond creating NaCl.

Since the electrostatic attraction increases as the ions get closer together, it might seem that equilibrium is not possible and the ions will collapse into one another. This is prevented by the Pauli exclusion principle discussed in Section 6.5. If the ions become too close, the core electrons begin to overlap spatially. Because of the exclusion principle, this can only happen if some electrons go into higher energy quantum states. This increase in energy when the ions are pushed closer together is equivalent to a repulsion of the ions. These two components, i.e. a long range attraction and a short range repulsion, can be combined to write the total potential energy in the form

$$U = -\frac{\alpha e^2}{4\pi\epsilon_0 r} + \frac{\beta}{r^\gamma}, \qquad (7.5)$$

where α, β and γ are constants: α is called the Madelung constant and is related to the geometry of the system and γ is called the Born exponent and is related to the compressibility of the solid. A sketch of the variation of potential energy with ionic separation is shown in Fig. 7.3.

The potential energy has its minimum value at the equilibrium separation, $r = r_0$, which can be found by solving

$$\left.\frac{\partial U}{\partial r}\right|_{r=r_0} = 0. \tag{7.6}$$

We can use this to rewrite (7.5) in terms of r_0 eliminating β in the process (this is set as an exercise at the end of the chapter) to give

$$U = -\frac{\alpha e^2}{4\pi\epsilon_0 r_0}\left(1 - \frac{1}{\gamma}\right). \tag{7.7}$$

This is called the Born–Landé equation and was worked out by Max Born and Alfred Landé in 1918 as a means of calculating the lattice energy of a crystalline ionic compound. Let us consider the example of NaCl. This has the values: $\gamma \approx 8$, $\alpha \approx 1.75$ and an equilibrium separation of the ions of $r_0 = 0.283$ nm. Substituting these values into the Born–Landé equation, we get

$$U \approx -7.8\,\text{eV}, \tag{7.8}$$

which is commonly written as $-751\,\text{kJ/mol}$. Given the simplicity of the model, this compares very well with the experimentally determined value for the lattice energy of NaCl, $U \approx -774\,\text{kJ/mol}$.

7.5. Covalent Bonding

A very different mechanism is responsible for the bonding of similar or identical atoms, such as gaseous hydrogen (H_2), nitrogen (N_2) and carbon monoxide (CO). If we calculate the energy required to create H^+ and H^- by transferring an electron between two hydrogen atoms and add to this the electrostatic potential energy, we find that there is no distance for which the total energy is negative. In other words, two hydrogen atoms cannot form an ionic bond. Instead, the attraction is a purely quantum effect due to the sharing of the two electrons by both atoms. This is called a covalent bond.

We can understand covalent bonding by considering the example of two finite square wells that are close to each other (see Fig. 7.4). These wells represents the attractive potential the electrons experience due to the two positively charged nuclei. We begin by considering a single electron that is equally likely to be found in either well. Because the wells are identical, the probability density, which is proportional to $|\psi|^2$, must be symmetric around the midpoint of the wells. This means that the wave function must be either symmetric or antisymmetric with respect to the wells, as shown in Fig. 7.4.

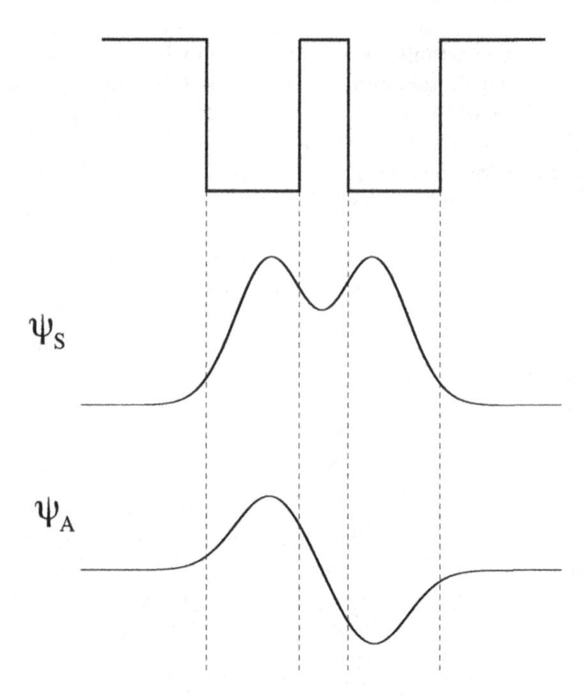

Fig. 7.4. Two square wells representing protons that are close together. Between the wells, the antisymmetric spatial wave function, ψ_A, of the electrons is close to zero, whereas the symmetric spatial wave function, ψ_S, is large.

Now, we consider the case of two electrons (i.e. one from each hydrogen atom). We know that electrons are fermions (with spin 1/2) and, therefore, the total wave function must be antisymmetric on exchange of the electrons. The total wave function is the product of a spatial part and a spin part. So the wave function of the two electrons can be a product of a symmetric spatial part and an antisymmetric spin part or of a symmetric spin part and an antisymmetric spatial part.

As can be seen in Fig. 7.4, the probability distribution, $|\psi|^2$ in the region between the protons is large for the spatially symmetric wave function and small for the spatially antisymmetric wave function. Thus when the space part of the wave function is symmetric and the electrons are in the $S = 0$ singlet state, the electrons are more likely to be found in the space between the protons, and the protons are bound together by this negatively charged cloud. Conversely, when the space part of the wave function is antisymmetric and the electrons are in the $S = 1$ triplet state, the electrons spend very little time between the protons and do not bind them together into a molecule. The symmetric spatial wave function, therefore, has a lower potential energy. The equilibrium separation for H_2 is $r = 0.074$ nm, and the binding energy is 4.52 eV. For the antisymmetric spatial state, the potential energy is never negative and so there is no bonding.

When two identical atoms are covalently bonded, symmetry dictates that the electron must be evenly shared between them. If, however, the atoms are different, the shared electrons will be more strongly attracted to one or the other. Such a bond is said to be polar and results in an unequal distribution of the electrons. This gives a permanent net electric dipole to the molecule. Polar bonds play an important role in determining the properties of molecules. They are also important for another type of bond that occurs *between* molecules – the van der Waals interaction.

7.6. Van der Waals Force

If we take a gas of atoms or molecules and cool it down, it will typically condense into a liquid. If we cool it even further, it will eventually solidify[4]. This suggests that there are attractive forces between the atoms or molecules that are different from the ones discussed above. At high temperatures, the thermal energy completely overwhelms these weak forces and we don't notice them. The particles are essentially unbound and we have a gas. However, as the thermal energy is reduced, the weak intermolecular forces start to become important – this is when the gas begins to condense. These weak interactions are called van der Waals forces.

There are actually three types of van der Waals forces. The first type is called the dipole-dipole interaction (or sometimes the Keesom interaction). It occurs between different molecules that have permanent dipoles arising from their polar bonds. Electrostatic forces are generated between the opposing charges of the dipoles and the molecules tend to align themselves to increase the attraction and reduce their energy. The second source of attraction is induction (also known as polarization) and occurs between a molecule with a permanent dipole and a molecule with no dipole. The molecule with a permanent dipole will influence the other molecule and cause a charge imbalance on it. This results in a dipole on the other molecule and then the two are attracted. The third type of interaction is called the London dispersion force[5]. Bizarrely, this force occurs between two molecules when neither of them has a dipole. It relies on the fact that electron densities fluctutate and so, although the molecules have no net dipole, they can have instantaneous charge imbalances. Just as was the case for induced van der Waals forces, these fleeting dipoles can induce equally fleeting opposite dipoles on a neighbouring molecule and the two are attracted. Not surprisingly, this dispersion force is very weak but it is, nonetheless, a very real effect.

Hydrogen bonds are a stronger form of dipole-dipole interactions, caused by highly electronegative atoms. They only occur between hydrogen and oxygen, fluorine or nitrogen, and are the strongest intermolecular force. The high electronegativities of F, O and N create highly polar bonds with hydrogen, which

[4]An exception is helium that never solidifies at atmospheric pressure.

[5]Named after the physicist Fritz London, not the city.

leads to strong bonding between hydrogen atoms on one molecule and the F, O or N atoms on adjacent molecules. Water (H_2O) is an important example of a polar molecule that experiences hydrogen bonding. Many of the special properties of water are due to the hydrogen bonds. For example, ice floats because hydrogen bonds hold water molecules further apart in a solid than in a liquid. Water's strong surface tension, high specific heat and widespread solvent properties are also due to hydrogen bonding.

Van der Waals forces can be conveniently modelled using the Lennard–Jones potential, which was proposed by John Lennard–Jones in 1924. It has the form

$$U = 4\epsilon \left[\left(\frac{\sigma}{r}\right)^{12} - \left(\frac{\sigma}{r}\right)^{6} \right], \tag{7.9}$$

where σ and ϵ are parameters that are specific to the molecules being considered. For example, for water $\sigma \approx 0.32$ nm and $\epsilon \approx 0.65$ kJ/mol. We see that it has a similar general form to the potential for ionic bonding (7.5) we saw earlier: there is a short range repulsion due to the Pauli exclusion principle and a longer range attractive force. One notable difference, however, is that the exponents are much larger. This means that the van der Waals force is quite short-ranged. The Lennard–Jones potential is just an approximation, but it has proven to be a very useful tool for studying the behaviour of interactions between molecules.

7.7. Lasers

One very important application that brings together the ideas of atomic structure, particle statistics and thermal physics is the laser[6]. In this section, we will discuss how lasers operate and derive the 'lasing condition', i.e. the criterion that needs to be satisfied for a laser to work.

The foundations for the discovery of the laser were set down by Einstein when he explained coherent and spontaneous atomic transitions in 1917. We know that an atom can absorb light of frequency f and make a transition from an energy state E_1 to E_2, where $E_2 - E_1 = hf$. This process is called absorption and the rate of absorption per atom can be written as $B_{12}U(f,T)$, where B_{12} is Einstein's coefficient of absorption and $U(f,T)$ is the energy density of the light field, which depends on the frequency, f, and the temperature, T. Once the atom is excited, it can make a *spontaneous* (i.e. not governed by the light field) transition to the lower level again. This process of spontaneous emission is accompanied by the emission of a photon of frequency, f, and the rate of transition per atom is given by the coefficient A_{21} or, equivalently, by the lifetime $t_s = 1/A_{21}$. In addition to these two processes, it is possible for the atom to undergo *stimulated emission*. This is an emission process that does depend on the light field. If an excited atom is illuminated with resonant light, it

[6]The laser is a classic example of the potential benefits of fundamental research. When it was first discovered it was viewed as an 'exotic flashlight' and a 'solution waiting for a problem'. Now its technological uses are vast, ranging from DVD players to cutting devices and barcode readers to surgical scalpels.

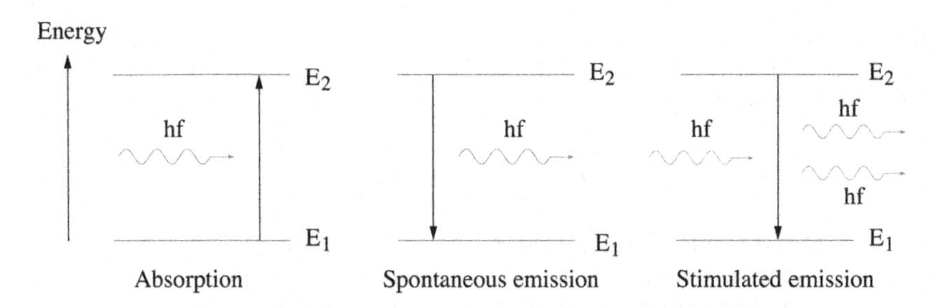

Fig. 7.5. The processes of (a) Absorption (b) Spontaneous emission and (c) Stimulated emission.

can undergo a transition to the lower level with the emission of two identical photons. These photons have the same frequency, are emitted in the same direction and are coherent (i.e. have the same phase or oscillate 'in step'). It is because photons are bosons that they are able to occupy the same quantum state. The rate of this process per atom can be written as $B_{21}U(f,T)$, where B_{21} is Einstein's coefficient of stimulated emission. These three processes are summarised in Fig. 7.5.

Now, if we have a collection of atoms and radiation in thermal equilibrium at temperature, T, the populations, N_1 and N_2 in levels E_1 and E_2 obey the Boltzmann distribution,

$$\frac{N_2}{N_1} = e^{-hf/kT}. \tag{7.10}$$

We can see this, for example, from Eq. (2.2). In the high temperature limit, which is appropriate for lasers, the energy for level i with frequency f_i is $E_i \approx hf_i e^{-hf_i/kT}$. So the ratio of populations in two levels (1 and 2) is just the ratio of the energy in each level divided by the energy per particle in that level, i.e.

$$\frac{N_2}{N_1} = \frac{E_2/hf_2}{E_1/hf_1} = \frac{e^{-hf_2/kT}}{e^{-hf_2/kT}} = e^{-hf/kT}, \tag{7.11}$$

where $f = f_2 - f_1$ is the difference in frequencies between the two levels. This ratio is maintained by dynamic equilibrium whereby the total rate of upward transitions is balanced by the total rate of downward transitions. This can be written as

$$N_1U(f,T)B_{12} = N_2B_{21}U(f,T) + N_2A_{21}, \tag{7.12}$$

where the left side is the rate of upward transitions (absorption) and the right side is the rate of downward transitions (stimulated and spontaneous emission). Using (7.10), this becomes,

$$U(f,T) = \frac{A_{21}}{B_{12}e^{hf/kT} - B_{21}}. \tag{7.13}$$

However, we already have a well-known expression for the energy density of a radiation field in thermal equilibrium with a set of oscillators: Planck's black body energy density that we met earlier in the book (see Eq. (2.4)),

$$U(f,T) = \frac{8\pi h f^3}{c^3} \frac{1}{e^{hf/kT} - 1}. \tag{7.14}$$

Comparing (7.13) and (7.14) we get $B_{21} = B_{12} \equiv B$ and $A_{21}/B = 8\pi h f^3/c^3$. The first equation shows that the rates of stimulated absorption and emission per atom must be the same. The second equation shows that at high frequencies $A_{21} \gg B$, i.e. spontaneous emission dominates.

It is useful to define a stimulated transition rate per atom (as we have seen, this is the same for emission or absorption)

$$W_i = U(f,T)B = \frac{U(f,T)c^3}{8\pi h f^3 t_s} \tag{7.15}$$

where the second part follows from writing B in terms of A_{21} and then substituting $A_{21} = 1/t_s$. This can be rewritten in terms of the intensity of the radiation field, $I(f,T) = U(f,T)c$,

$$W_i = \frac{c^2}{8\pi h f^3 t_s} g(f) I(f,T), \tag{7.16}$$

where we have also included the atomic line width $g(f)$, which accounts for the finite energy widths of the levels[7].

7.8. The Lasing Condition

We have seen that rates of stimulated absorption and emission are the same per atom. When light is incident on a collection of atoms, there is usually a net absorption of light because, due to the Boltzmann distribution, there are more atoms in the ground state than the excited state. However, if were possible to put more atoms in the excited state than the ground state (a condition known as *population inversion*), there would be an amplification of the photons, i.e. a laser[8].

Laser light has the useful properties of coherence, monochromaticity, high intensity and low beam divergence and lasers all have the following key features:

1. An energy source capable of producing population inversions. The process of providing this energy is called pumping and it is typically supplied as an electrical current or as light using a flash lamp or perhaps another laser.

[7]This atomic line width generally has the shape of a Lorentzian

$$g(f) = \frac{\Delta f}{2\pi\left[(f - f_c)^2 + (\Delta f/2)^2\right]} \tag{7.17}$$

centered at frequency f_c and with a full width at half maximum of Δf.

[8]Laser is an acronym for light amplification by stimulated emission of radiation.

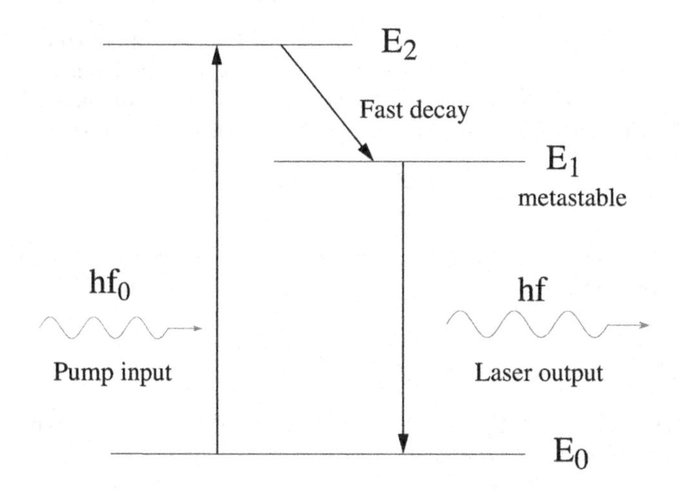

Fig. 7.6. The energy levels of a three-level atom capable of supporting a population inversion.

2. A lasing medium (e.g. ruby, argon, carbon dioxide or a mixture of helium and neon) with energy levels that are capable of supporting a population inversion.

3. A method of containing the emitted photons so that they can stimulate further emission from other excited atoms. This is usually achieved by placing mirrors at each end of the lasing medium so that photons make multiple passes through it. The laser beam is extracted by ensuring that one of the mirrors is partially transmitting.

In order to achieve a population inversion we require a lasing medium with at least three energy levels. To see this suppose we only had two levels. Boltzmann's distribution tells us that there are initially more atoms in the lower level. If we now pump the system, there will initially be a net transfer of atoms to the upper level. Eventually, if we pump very hard and so can ignore spontaneous emission, we will reach the point where the upper and lower levels have the same population. What happens if we keep pumping? We know that the upward and downward transition rates will be exactly balanced and so there is no further net transfer of population and, therefore, we cannot get a population inversion with two levels. One possible solution is to use a three-level system: a ground state, an intermediate state with a relatively long lifetime t_s and a high energy pump state (see Fig. 7.6). Now as we pump atoms from the ground state to the highest state, they quickly decay into the metastable intermediate state before they can get back to the ground state by stimulated or spontaneous emission. This enables us to achieve a population inversion between the ground and intermediate states.

Laser action will be sustained if the rate of increase of coherent photons is larger than the rate of decrease due to various loss mechanisms, such as transmission at the mirrors and scattering processes. If we combine all the loss mechanisms into a single time constant, t_p, the intensity will damp exponentially as

$$\left(\frac{dI}{dt}\right)_{\text{loss}} = -\frac{I}{t_p}. \tag{7.18}$$

The gain in coherent intensity is given by the sum of the gain by stimulated emission from level 2 and the loss due to coherent absorption from level 1

$$\left(\frac{dI}{dt}\right)_{\text{gain}} = \Delta N \frac{hfc}{V} W_i, \tag{7.19}$$

where $\Delta N = N_2 - N_1$, V is the laser volume and hfc/V is the intensity contributed per photon. For sustained laser action we require

$$\left(\frac{dI}{dt}\right)_{\text{gain}} \geq \left|\left(\frac{dI}{dt}\right)_{\text{loss}}\right|, \tag{7.20}$$

which gives

$$\frac{\Delta N}{V} hfcW_i \geq \frac{I}{t_p}. \tag{7.21}$$

Substituting from (7.16), the critical population inversion density is

$$\frac{\Delta N_c}{V} = \frac{8\pi f^2 t_s}{c^3 g(f) t_p}. \tag{7.22}$$

For a Lorentzian line width (see Eq. (7.17)) centred on f_c, we have $g(f_c) = 2/(\pi\Delta f)$. Substituting this into (7.22), we get

$$\frac{\Delta N_c}{V} = \frac{4\pi^2 f^2 \Delta f}{c^3} \left(\frac{t_s}{t_p}\right). \tag{7.23}$$

This is a particularly useful equation because it tells us the required population inversion density for laser action with a transition of known frequency, width and lifetime.

7.9. Exercises

1. What is the total number of quantum states of hydrogen with quantum number $n = 4$?

2. Taking into account the spin, show that the degeneracy of an energy level of the hydrogen atom with principal quantum number n is given by $2n^2$. You might find the following result useful:

$$\sum_{k=1}^{N} k = \frac{N(N+1)}{2}.$$

3. If the electron configuration of an atom is $1s^2 2s^2 2p^6 3s^2 3p^6 4s^2$, what element is the atom?

4. What is the longest wavelength in the absorption spectrum of hydrogen (initially in the ground state)? What is the transition?

5. Determine whether each of the following electron configurations represents a possible state of an element. If it is possible then determine whether it is the ground state or an excited state.
(a) $1s^2 2s^2 2p^4 3s^2$
(b) $1s^2 2s^6 2p^6 3s$
(c) $1s^2 2s^2 2p^6 3s^2 3p^2$
(d) $1s^2 2s^2 2p^1$
(e) $1s^1 2s^2 2p^6 3s^2$

6. The potential energy of a pair of ions in an ionic solid as a function of their separation, r, is given by

$$U(r) = -\frac{\alpha e^2}{4\pi\epsilon_0 r} + \frac{\beta}{r^\gamma},$$

where α, β and γ are constants. Find an expression for the minimum potential energy in terms of the equilibrium separation, r_0.

7. Using energy considerations, find an upper bound to the ionic bond length for NaCl.

8. Show that the force on an ion in an ionic bond can be written as

$$F = -\frac{\alpha e^2}{4\pi\epsilon_0 r^2}\left[1 - \left(\frac{r_0}{r}\right)^{\gamma-1}\right].$$

Using this expression, show that if one of the ions is displaced from its equilibrium position, r_0 by some small amount $x \ll r_0$, then the restoring force can be approximated by Hooke's law. In other words the force can be written in the form

$$F = -Kx,$$

where K is a constant.

Chapter 8

Formal Structure of Quantum Mechanics

In the preceding chapters, we have introduced some important ideas in quantum mechanics and used them to study a number of physical systems. The formulation of the theory we used is known as wave mechanics and is due largely to Schrödinger. We are now going to take a more formal approach and will see that quantum mechanics can be formulated in different ways, such as Heisenberg's matrix mechanics, that are equivalent to wave mechanics. We will also introduce the very useful Dirac notation, which is widely used by researchers and can simplify calculations a lot. First let us introduce the important ideas of states and ensembles.

8.1. States and Ensembles

The *state* of a system is a description of all the properties that can be known about it. Another way of saying this is that the state contains all the information you would require to make an identical copy of a system. Classically, the state of a system would include such things as the position and velocity of every particle. In quantum mechanics, the state could, for example, be expressed by the wave function or by listing the values of all the relevant quantum numbers. Although there are different ways of writing the state, the information it contains is the same.

Quantum mechanics is fundamentally different from classical physics in that it is probabilistic. Measurements made on identically prepared systems do not necessarily result in the same measurement outcome, which means that to characterise a state (e.g. determine the probability of finding a particle at a certain position) we need to perform many measurements on identical systems. A similar idea arises in random classical processes. Suppose we wanted to determine if a coin was 'fair'. Tossing it once would not allow us to do that. We would need to make many measurements to determine the probability of heads versus tails.

Of course, with a classical system, we could use the same coin over and over, whereas with a quantum system, the act of measurement changes the system we are observing and so we need a collection of identical systems. Such a collection of N identically prepared systems (where N is large) is called an *ensemble* and is a very important concept in quantum physics.

Suppose now that we have a system that has a set of measurement outcomes $\{\alpha_1, \alpha_2, \alpha_3, ...\}$ with corresponding probabilities P_1, P_2, P_3,...[1] The expectation value of the measurement on an ensemble is given by

$$\langle \alpha \rangle = \sum_i \alpha_i P_i.$$

Two physical states are identical if the probabilities are the same for all possible measurement outcomes, i.e. there is no possible measurement that can tell them apart.

8.2. Introduction to Dirac Notation

Dirac notation is a convenient tool used to represent the quantum mechanical state of a system. It is just another way of writing vectors. Since quantum mechanics is concerned with how states (or vectors) are transformed by physical processes, it relies heavily on linear algebra. Dirac notation greatly simplifies these calculations.

The Dirac notation is made up of bras $\langle . |$ and kets $| . \rangle$, which together form bra(c)kets, $\langle . || . \rangle$, usually just written as $\langle . | . \rangle$. The state of a system is represented by a ket and can be written as a column vector. The corresponding bra is the adjoint of the ket, i.e. the complex conjugate and transpose, and can be written as a row vector. Consider, for example, the case of an electron. We have seen that it has two possible basis states: spin up and spin down. In general, the electron can be in a superposition of these two possibilities and so the state in Dirac notation is

$$|\psi\rangle = \alpha|\uparrow\rangle + \beta|\downarrow\rangle, \tag{8.1}$$

which can be written as the equivalent column vector

$$|\psi\rangle = \begin{pmatrix} \alpha \\ \beta \end{pmatrix}. \tag{8.2}$$

The corresponding bra can be written as

$$\langle\psi| = \alpha^*\langle\uparrow| + \beta^*\langle\downarrow| \qquad \text{or} \qquad \langle\psi| = (\alpha^* \;\; \beta^*). \tag{8.3}$$

Often in quantum mechanics we want to calculate the overlap of one state $|\psi\rangle$ with another $|\phi\rangle$. This is found simply by calculating the bra-ket of the two

[1] We are keeping the formalism general at this stage, but the measurements could, for example, correspond to the different energy levels in the quantum harmonic oscillator.

states, i.e. $\langle\phi|\psi\rangle$, and is called the inner product. The overlap of a state, $|\psi\rangle$ with itself is called the norm of $|\psi\rangle$ and is given by

$$\langle\psi|\psi\rangle = (\alpha^* \quad \beta^*)\begin{pmatrix}\alpha \\ \beta\end{pmatrix} = |\alpha|^2 + |\beta|^2. \tag{8.4}$$

Normalisation corresponds to setting the norm equal to unity, i.e. $|\alpha|^2+|\beta|^2 = 1$. We can also work the norm out directly from bra and ket notation by noting

$$\langle\uparrow\mid\downarrow\rangle = \langle\downarrow\mid\uparrow\rangle = 0 \tag{8.5}$$

$$\langle\uparrow\mid\uparrow\rangle = \langle\downarrow\mid\downarrow\rangle = 1. \tag{8.6}$$

These relations mean that the basis states are normalised and are orthogonal to each other (i.e. their inner product with each other is zero). Writing out the expression for the norm in bra-ket notation, we get

$$\begin{aligned}\langle\psi|\psi\rangle &= (\alpha^*\langle\uparrow\mid + \beta^*\langle\downarrow\mid)(\alpha\mid\uparrow\rangle + \beta\mid\downarrow\rangle) \\ &= \alpha^*\alpha\langle\uparrow\mid\uparrow\rangle + \beta^*\beta\langle\downarrow\mid\downarrow\rangle + \alpha^*\beta\langle\uparrow\mid\downarrow\rangle + \beta^*\alpha\langle\downarrow\mid\uparrow\rangle \\ &= |\alpha|^2 + |\beta|^2,\end{aligned}$$

which is the same as we calculated above.

So far, we have represented the state in one particular basis: spin-up and spin-down. This is a natural way of writing the state of a spin-half particle, but it is not the only way. In fact, there are an infinite number of equally good bases. We could, for example, choose to write our state (8.1) in the basis of the symmetric and antisymmetric combinations of $\mid\uparrow\rangle$ and $\mid\downarrow\rangle$, i.e.

$$|+\rangle = \frac{1}{\sqrt{2}}(\mid\uparrow\rangle + \mid\downarrow\rangle) \tag{8.7}$$

$$|-\rangle = \frac{1}{\sqrt{2}}(\mid\uparrow\rangle - \mid\downarrow\rangle). \tag{8.8}$$

These new basis states obey similar rules to before

$$\langle+|-\rangle = \langle-|+\rangle = 0 \tag{8.9}$$

$$\langle+|+\rangle = \langle-|-\rangle = 1. \tag{8.10}$$

Using this new basis, the state $|\psi\rangle = \alpha\mid\uparrow\rangle + \beta\mid\downarrow\rangle$ can be written as

$$|\psi\rangle = \frac{\alpha+\beta}{\sqrt{2}}|+\rangle + \frac{\alpha-\beta}{\sqrt{2}}|-\rangle, \tag{8.11}$$

which is easily confirmed by substitution.

We can generalise our results for two states to systems with a higher number of dimensions, e.g. the states corresponding to different energy levels in the quantum harmonic oscillator. If we have n states and label them as $\{0, 1, ..., n-1\}$, a general state in this basis can be written as

$$|\psi\rangle = \alpha_0|0\rangle + \alpha_1|1\rangle + \cdots + \alpha_{n-1}|n-1\rangle \tag{8.12}$$

$$= \sum_{i=0}^{n-1}\alpha_i|i\rangle, \tag{8.13}$$

where $\{\alpha_i\}$ are complex numbers. The equivalent column vector representation of this state is

$$|\psi\rangle = \begin{pmatrix} \alpha_0 \\ \vdots \\ \alpha_{n-1} \end{pmatrix}. \tag{8.14}$$

As before, we can define

$$\langle i|j\rangle = \begin{Bmatrix} 0 & \text{if } i \neq j \\ 1 & \text{if } i = j \end{Bmatrix} = \delta_{i,j}. \tag{8.15}$$

Physically, the orthogonality of these states means that they can be fully distinguished in principle.

Now suppose that we have another state,

$$|\phi\rangle = \sum_{i=0}^{n-1} \beta_i |i\rangle, \tag{8.16}$$

where $\{\beta_i\}$ are complex numbers. The inner product of (8.13) and (8.16) is then

$$\langle\phi|\psi\rangle = \left(\sum_{i=0}^{n-1} \beta_i^* \langle i|\right)\left(\sum_{j=0}^{n-1} \alpha_j |j\rangle\right) \tag{8.17}$$

$$= \sum_{i=0}^{n-1}\sum_{j=0}^{n-1} \beta_i^* \alpha_j \, \delta_{i,j} \tag{8.18}$$

$$= \sum_{i=0}^{n-1} \beta_i^* \alpha_i. \tag{8.19}$$

8.3. Operators

We have seen how to represent quantum states in Dirac notation. A large part of quantum mechanics is involved with studying how these states change due, for example, to time evolution or measurement. In this formalism, a quantum state is changed by an operator. You have already seen the position and momentum operators in the wave mechanics formalism in Section 3.5. In this section, we want to study operators in more detail and see how they can be represented in Dirac notation.

The simplest operator of all is the identity operator, because it does nothing. If you operate on any state with the identity operator, you get exactly the same state back. It is a useful starting point for understanding operators and is actually much more useful than it might appear at first sight, as we shall see later. Suppose we have a complete set of orthogonal and normalised basis states $\{|i\rangle\}_{i=0}^{n-1}$. The identity operator can then be written in Dirac notation as

$$I = \sum_{i=0}^{n-1} |i\rangle\langle i|. \tag{8.20}$$

When applied to the general state given by Eq. (8.13), we get,

$$I|\psi\rangle = \left(\sum_{i=0}^{n-1} |i\rangle\langle i|\right)\left(\sum_{j=0}^{n-1} \alpha_j\,|j\rangle\right) \tag{8.21}$$

$$= \sum_{i=0}^{n-1} \alpha_i|i\rangle = |\psi\rangle. \tag{8.22}$$

We see that the identity operator leaves the state unchanged.

Returning to the example of a spin-half particle, a general state can be written as a superposition of the spin-up and spin-down basis states. The identity operator can, therefore, be written as

$$I = |\uparrow\rangle\langle\uparrow| + |\downarrow\rangle\langle\downarrow|. \tag{8.23}$$

However, as we have seen, another perfectly good orthogonal and normalised basis is given by the basis states $|+\rangle$ and $|-\rangle$ defined by equations (8.7) and (8.8). This means that an alternative way of writing the identity operator is

$$I = |+\rangle\langle+| + |-\rangle\langle-| \tag{8.24}$$

A useful application of the identity operator is to change the representation of the state vector. We can demonstrate this with the example given in Section 8.2. Suppose we want to write the state given by (8.1) in the $\{|+\rangle, |-\rangle\}$ basis. We do not want to change the state – just it's *representation*. We know that the identity operator leave the state unchanged, so this suggests a way of changing the basis. We operate on the state (written in the $\{|\uparrow\rangle, |\downarrow\rangle\}$ basis) with the identity operator (written in the $\{|+\rangle, |-\rangle\}$ basis). This gives

$$I|\psi\rangle = (|+\rangle\langle+| + |-\rangle\langle-|)(\alpha|\uparrow\rangle + \beta|\downarrow\rangle) \tag{8.25}$$

$$= \alpha|+\rangle\langle+|\uparrow\rangle + \beta|+\rangle\langle+|\downarrow\rangle + \alpha|-\rangle\langle-|\uparrow\rangle + \beta|-\rangle\langle-|\downarrow\rangle \tag{8.26}$$

$$= (\alpha\langle+|\uparrow\rangle + \beta\langle+|\downarrow\rangle)|+\rangle + (\alpha\langle-|\uparrow\rangle + \beta\langle-|\downarrow\rangle)|-\rangle. \tag{8.27}$$

Now, calculating the inner products contained within round brackets and substituting back: $\langle+|\uparrow\rangle = \langle-|\uparrow\rangle = \langle+|\downarrow\rangle = 1/\sqrt{2}$ and $\langle-|\downarrow\rangle = -1/\sqrt{2}$, we get

$$|\psi\rangle = \frac{\alpha+\beta}{\sqrt{2}}|+\rangle + \frac{\alpha-\beta}{\sqrt{2}}|-\rangle, \tag{8.28}$$

which is the transformed state we saw earlier.

In general, operators transform states and, since states are represented by vectors, operators must be represented by matrices. A general matrix for an operator A in two-dimensional state space, e.g. spin up and spin down, is

$$A = \begin{pmatrix} a_1 & a_2 \\ a_3 & a_4 \end{pmatrix}. \tag{8.29}$$

This can be written in Dirac notation using the outer product of bras and kets

$$A = a_1|\uparrow\rangle\langle\uparrow| + a_2|\uparrow\rangle\langle\downarrow| + a_3|\downarrow\rangle\langle\uparrow| + a_4|\downarrow\rangle\langle\downarrow|. \quad (8.30)$$

We can calculate the effect of this operator on a general state $|\psi\rangle = \alpha|\uparrow\rangle + \beta|\downarrow\rangle$

$$\begin{aligned} A|\psi\rangle &= (a_1|\uparrow\rangle\langle\uparrow| + a_2|\uparrow\rangle\langle\downarrow| + a_3|\downarrow\rangle\langle\uparrow| + a_4|\downarrow\rangle\langle\downarrow|)(\alpha|\uparrow\rangle + \beta|\downarrow\rangle) \\ &= (\alpha a_1 + \beta a_2)|\uparrow\rangle + (\alpha a_3 + \beta a_4)|\downarrow\rangle. \end{aligned} \quad (8.31)$$

The same calculation can be performed using matrix and vector notation,

$$A|\psi\rangle = \begin{pmatrix} a_1 & a_2 \\ a_3 & a_4 \end{pmatrix} \begin{pmatrix} \alpha \\ \beta \end{pmatrix} = \begin{pmatrix} \alpha a_1 + \beta a_2 \\ \alpha a_3 + \beta a_4 \end{pmatrix} \quad (8.32)$$

and we see that we get the same result as in (8.31).

As another example, consider the identity operator. We know that the identity matrix has the form,

$$I = \begin{pmatrix} 1 & 0 \\ 0 & 1 \end{pmatrix}. \quad (8.33)$$

Comparing with (8.29) and (8.29), we get $a_1 = a_4 = 1$ and $a_2 = a_3 = 0$ and so

$$I = |\uparrow\rangle\langle\uparrow| + |\downarrow\rangle\langle\downarrow|, \quad (8.34)$$

which is the form of the identity operator we saw earlier in Eq. (8.23).

8.4. Measurements

When measurements are made on quantum mechanical systems, the state of the system is changed. In particular, it is *projected* onto the measurement outcome. This is sometimes referred to as the collapse of the wave function. The operator that performs this process is called a projector and has the general form

$$P_\phi = |\phi\rangle\langle\phi|. \quad (8.35)$$

This projects a general state vector onto $|\phi\rangle$ as can be seen by direct calculation

$$P|\psi\rangle = |\phi\rangle\langle\phi|\psi\rangle, \quad (8.36)$$

where $\langle\phi|\psi\rangle$ is just a complex number. If we normalise the state, we are left with $|\phi\rangle$. It is worth noting that applying a projector twice is the same as applying it once, i.e.

$$P_\phi P_\phi = |\phi\rangle\langle\phi|\phi\rangle\langle\phi| = |\phi\rangle\langle\phi| = P_\phi. \quad (8.37)$$

This is what we would expect because if we project with P_ϕ, we are left with the state $|\phi\rangle$ and so any further projection into $|\phi\rangle$ will not change anything.

Let us now consider measurements on an ensemble of identically prepared systems to see how expectation values can be written in Dirac notation. Taking the example of a spin-half particle in some general state $|\psi\rangle$, a measurement of the particle as spin-up corresponds to a value of $S_z = \hbar/2$ and measuring the particle as spin-down corresponds to a value of $S_z = -\hbar/2$. The expectation value for the z-component of the spin is therefore

$$\langle S_z\rangle = \frac{\hbar}{2}|\langle\uparrow|\psi\rangle|^2 - \frac{\hbar}{2}|\langle\downarrow|\psi\rangle|^2, \tag{8.38}$$

where $|\langle\uparrow|\psi\rangle|^2$ and $|\langle\downarrow|\psi\rangle|^2$ are respectively the probabilities of finding the particle in spin-up and spin-down. Rearranging, we get

$$\langle S_z\rangle = \frac{\hbar}{2}\langle\psi|\uparrow\rangle\langle\uparrow|\psi\rangle - \frac{\hbar}{2}\langle\psi|\downarrow\rangle\langle\downarrow|\psi\rangle \tag{8.39}$$

$$= \langle\psi|\left[\frac{\hbar}{2}|\uparrow\rangle\langle\uparrow| - \frac{\hbar}{2}|\downarrow\rangle\langle\downarrow|\right]|\psi\rangle \tag{8.40}$$

$$= \langle\psi|\hat{S}_z|\psi\rangle, \tag{8.41}$$

where the spin operator is given by

$$\hat{S}_z = \frac{\hbar}{2}\left[|\uparrow\rangle\langle\uparrow| - |\downarrow\rangle\langle\downarrow|\right], \tag{8.42}$$

or equivalently

$$\hat{S}_z = \frac{\hbar}{2}\begin{pmatrix} 1 & 0 \\ 0 & -1 \end{pmatrix}. \tag{8.43}$$

We see from (8.41) that the expectation value has a particularly simple form in Dirac notation. In general, the expectation value for any observable with operator $\hat{A}$ for an ensemble of systems in state $|\psi\rangle$ is

$$\langle A\rangle = \langle\psi|\hat{A}|\psi\rangle. \tag{8.44}$$

Let us now take a closer look at how operators are constructed. For this, it is useful to revisit eigenstates and eigenvalues in the context of Dirac notation. An eigenstate of an operator $\hat{A}$ is any vector $|e\rangle$ such that:

$$\hat{A}|e\rangle = \lambda|e\rangle \tag{8.45}$$

and λ is the eigenvalue associated with the eigenvector $|e\rangle$. For example, $|\uparrow\rangle$ and $|\downarrow\rangle$ are the eigenstates of the operator $\hat{S}_z$ and the corresponding eigenvalues are $+\hbar/2$ and $-\hbar/2$ respectively. We can see this because

$$\hat{S}_z|\uparrow\rangle = +\frac{\hbar}{2}|\uparrow\rangle, \tag{8.46}$$

$$\hat{S}_z|\downarrow\rangle = -\frac{\hbar}{2}|\downarrow\rangle. \tag{8.47}$$

For all operators in quantum mechanics that represent physical observables, there exists a basis such that they can be written as

$$\hat{A} = \sum_{i=0}^{n-1} \alpha_i |i\rangle\langle i|, \tag{8.48}$$

where the basis states are all orthogonal and normalised. From this form of the operator, it is straightforward to show that $\{|i\rangle\}$ are the eigenstates of $\hat{A}$, with corresponding eigenvalues $\{\alpha_i\}$, since

$$\hat{A}|j\rangle = \sum_{i=0}^{n-1} \alpha_i |i\rangle\langle i|j\rangle \tag{8.49}$$

$$= \sum_{i=0}^{n-1} \alpha_i |i\rangle \delta_{i,j} = \alpha_j |j\rangle. \tag{8.50}$$

We can use the form of $\hat{A}$ given by (8.48) to calculate expectation values. A general state vector in an n-dimensional space has the form

$$|\psi\rangle = \sum_{i=0}^{n-1} c_i |i\rangle, \tag{8.51}$$

and so the expectation value of A on an ensemble of measurements is

$$\langle A\rangle = \langle\psi|\hat{A}|\psi\rangle \tag{8.52}$$

$$= \sum_{i=0}^{n-1} c_i^* \langle i| \left(\sum_{j=0}^{n-1} \alpha_j |j\rangle\langle j| \right) \sum_{k=0}^{n-1} c_k |k\rangle \tag{8.53}$$

$$= \sum_{i=0}^{n-1}\sum_{j=0}^{n-1}\sum_{k=0}^{n-1} c_i^* c_k \alpha_j \langle i|j\rangle\langle j|k\rangle \tag{8.54}$$

$$= \sum_{i=0}^{n-1}\sum_{j=0}^{n-1}\sum_{k=0}^{n-1} c_i^* c_k \alpha_j \delta_{i,j}\delta_{j,k} \tag{8.55}$$

$$= \sum_{i=0}^{n-1} |c_i|^2 \alpha_i \tag{8.56}$$

as we would expect.

8.5. Postulates of Quantum Mechanics

We can now restate the the fundamental postulates of quantum mechanics that we encountered at the start of Chapter 3, but this time using Dirac notation. These postulates, which are equivalent to the ones we saw earlier, tell us all we need to know about quantum systems by specifying what we mean by states, observables, measurements and evolution. The postulates can be stated as:

1. **States:** The state of a (pure) quantum system can be represented by a vector, $|\psi\rangle$, of unit length.

2. **Observables:** Quantum mechanical observables are represented by operators of the form

$$\hat{A} = \sum_{i=0}^{n-1} \alpha_i |i\rangle\langle i|, \tag{8.57}$$

   where $\{\alpha_i\}$ are the (real) possible measurement outcomes and $\{|i\rangle\}$ are the eigenstates of $\hat{A}$.

3. **Measurements:** A quantum state, $|\psi\rangle$, can be measured by projection onto a set of orthogonal states. If $|\phi_1\rangle, \cdots |\phi_n\rangle$ is such a set of orthogonal (and normalised) states, the state $|\psi\rangle$ will collapse onto $|\phi_i\rangle$ with probability $|\langle\phi_i|\psi\rangle|^2$.

4. **Unitary Evolution:** Any change in a quantum state that is not due to a measurement can be expressed by a unitary operator. Such an operator does not change the normalisation of the state. The Schrödinger equation, for example, evolves a state in a unitary fashion.

8.6. Position and Momentum Operators

Let us now introduce the operators, which play a major role when describing a quantum mechanical particle, like an electron, in free space, trapped in a square well, a harmonic potential or within the trapping potential of the nuclei of the hydrogen atom. These are the position operator $\hat{x}$, the momentum operator $\hat{p}$ and functions of them.

The possible measurement outcomes of a position measurement of a one dimensional particle are the real numbers between $-\infty$ and ∞. Moreover, we need to know the state vectors that yield a certain measurement outcome x with unit probability. Then we can construct the operator $\hat{x}$, as suggested by the second postulate of quantum mechanics. In the following we denote the states with a well defined position by $|x\rangle$. This allows us to write $\hat{x}$ as

$$\hat{x} = \int_{-\infty}^{\infty} dx\, x\, |x\rangle\langle x|. \tag{8.58}$$

The states $|x\rangle$ have to be normalised and pairwise orthogonal.

For completeness, we now introduce the momentum operator $\hat{p}$. The possible measurement outcomes of a momentum measurement on a quantum mechanical particle in one dimension are all possible numbers p. In the following, the state $|p\rangle$ denotes a state with the well-defined momentum p. Consequently, the momentum operator $\hat{p}$ can be written as

$$\hat{p} = \int_{-\infty}^{\infty} dp\, p\, |p\rangle\langle p|. \tag{8.59}$$

We will learn more about how to use the operators $\hat{x}$ and $\hat{p}$ later in the book.

A very useful principle in constructing quantum mechanical operators is the *correspondence principle.* It can be stated simply as follows: the operator of an observable $A = f(x, p)$ is given by $\hat{A} = f(\hat{x}, \hat{p})$. Two examples are:

1. The operator for a measurement of the momentum of a particle squared, i.e. for $A = p^2$, is given by

$$\hat{p}^2 = \int_{-\infty}^{\infty} dp\, p^2 |p\rangle\langle p| = (\hat{p})^2. \tag{8.60}$$

   This is consistent with the second postulate of quantum mechanics, since the possible measurement outcomes of a measurement of $\hat{p}^2$ are all possible p^2 and the states $|p\rangle$ are the states with a very well defined squared momentum p^2 .

2. The classical energy of a particle trapped in a potential $V(x)$ is given by $E = p^2/2M + V(x)$. Using the correspondence principle, the energy operator (also known as the Hamiltonian) is given by

$$\hat{H} = \frac{\hat{p}^2}{2M} + V(\hat{x}) \tag{8.61}$$

   with

$$V(\hat{x}) = \int_{-\infty}^{\infty} dx\, V(x)|x\rangle\langle x|. \tag{8.62}$$

   We will see shortly how this is important for the Schrödinger equation in Dirac notation.

One other result that is useful for constructing operators is the spectral theorem. Suppose some operator $\hat{A}$ can be written as

$$\hat{A} = \sum_{i=0}^{n-1} \lambda_i |\lambda_i\rangle\langle\lambda_i|, \tag{8.63}$$

by the second postulate. Then the function $f(\hat{A})$ of this operator is given by

$$f(\hat{A}) = \sum_{i=0}^{n-1} f(\lambda_i)|\lambda_i\rangle\langle\lambda_i|. \tag{8.64}$$

8.7. Position and Momentum Wave Functions

We have seen that quantum mechanics can be expressed in different ways. In the first part of the book, we studied different systems using wave mechanics. We have now introduced a new formalism based on vectors and matrices and expressed using Dirac notation. Of course these are just saying the same thing but in different ways and it is worth spending some time drawing the connection between them.

Suppose that we have some general quantum state $|\psi\rangle$ and that we want to express it in the position basis. We can do this by operating on it with the identity operator in the position basis, i.e.

$$|\psi\rangle = I|\psi\rangle = \int_{-\infty}^{\infty} dx\,|x\rangle\langle x|\psi\rangle \tag{8.65}$$

$$= \int_{-\infty}^{\infty} dx\,\psi(x)\,|x\rangle, \tag{8.66}$$

where $\psi(x) = \langle x|\psi\rangle$ is the wave function in position space that we studied earlier. It is nothing more than the amplitude of the general state $|\psi\rangle$ being found at position x, which is just how we interpreted the wave function earlier in the book (using the Born interpretation). This then draws the connection between matrix and wave mechanics.

In a similar way, we can express the state $|\psi\rangle$ in the momentum basis. This gives

$$|\psi\rangle = I|\psi\rangle = \int_{-\infty}^{\infty} dp\,|p\rangle\langle p|\psi\rangle \tag{8.67}$$

$$= \int_{-\infty}^{\infty} dp\,\tilde{\psi}(p)\,|p\rangle, \tag{8.68}$$

where $\tilde{\psi}(p)$ is the wave function of the state in momentum space.

8.8. Fourier Transforms and the Delta Function

Before we discuss the connection between $\psi(x)$ and $\tilde{\psi}(p)$, it is helpful to first take a short mathematical diversion to introduce two concepts that we will require: the delta function $\delta(x)$ and the Fourier transform.

The δ-function:
Consider a set of functions defined by

$$f_n(x) = \begin{cases} n & \text{for } -\frac{1}{2n} < x < \frac{1}{2n} \\ 0 & \text{otherwise.} \end{cases} \tag{8.69}$$

Then one can show that,

$$\int_{-\infty}^{\infty} dx\, f_n(x) = \frac{1}{n}n = 1 \tag{8.70}$$

for all n. The delta function $\delta(x)$ is defined as

$$\delta(x) = \lim_{n\to\infty} f_n(x) \tag{8.71}$$

and has the following properties:

$$\int_{-\infty}^{\infty} dx\, \delta(x) = 1 \tag{8.72}$$

$$\delta(x) = \begin{cases} \infty & \text{for } x = 0 \\ 0 & \text{for } x \neq 0 \end{cases} \tag{8.73}$$

$$\int_{-\infty}^{\infty} dx\, f(x)\delta(x - x_0) = \int_{-\infty}^{\infty} dy\, f(y + x_0)\delta(y) = f(x_0). \tag{8.74}$$

This last property is particularly useful because it means that the delta function can be used to 'pick out' the value of the $f(x)$ at a certain point, $x = x_0$.

The Fourier transform:
The function $\tilde{f}(k)$ is the Fourier transform of $f(x)$ if

$$\tilde{f}(k) = \frac{1}{\sqrt{2\pi}} \int_{-\infty}^{\infty} dx\, e^{-ikx} f(x). \tag{8.75}$$

As an example, the Fourier transform of the delta function, $f(x) = \delta(x)$, is

$$\tilde{f}(k) = \frac{1}{\sqrt{2\pi}} \int_{-\infty}^{\infty} dx\, e^{-ikx} \delta(x) = \frac{1}{\sqrt{2\pi}}, \tag{8.76}$$

i.e. the Fourier transform of a delta function is a completely flat function. The inverse transformation, which undoes the Fourier transform of a function, is

$$f(x) = \frac{1}{\sqrt{2\pi}} \int_{-\infty}^{\infty} dk\, e^{ikx} \tilde{f}(k). \tag{8.77}$$

We can show that these are inverses by directly showing that the inverse Fourier transform of the Fourier transform of a function gives the original function back. The Fourier transform of $f(x)$ is

$$\frac{1}{\sqrt{2\pi}} \int_{-\infty}^{\infty} dx'\, e^{-ikx'} f(x'), \tag{8.78}$$

and the inverse Fourier transform of this function is

$$\begin{aligned} \frac{1}{2\pi} \int_{-\infty}^{\infty} dk\, e^{ikx} \int_{-\infty}^{\infty} dx'\, e^{-ikx'} f(x') &= \frac{1}{2\pi} \int_{-\infty}^{\infty} dx'\, f(x') \int_{-\infty}^{\infty} dk\, e^{ik(x-x')} \\ &= \int_{-\infty}^{\infty} dx'\, f(x')\delta(x - x') \\ &= f(x), \end{aligned} \tag{8.79}$$

where we have made use of the identity,

$$\int_{-\infty}^{\infty} dk\, e^{ik(x-x')} = 2\pi\delta(x - x'). \tag{8.80}$$

This shows that (8.75) and (8.77) are inverses.

We are now in a position to derive a relationship between the position and momentum representations of the wave function. Suppose that the state vectors given by (8.66) and (8.68) represent the same state of a physical system, then there must be some fixed relationship between $\psi(x)$ and $\tilde{\psi}(p)$. Before we find the relationship, it is worth noting that $\langle x|p\rangle$ is the amplitude of a particle with well-defined momentum p being found at position x. A particle with fixed momentum $p = \hbar k$ is just a plane wave and so we can write

$$\langle x|p\rangle = Ae^{ipx/\hbar}, \tag{8.81}$$

where A is a constant. Similarly,

$$\langle p|x\rangle = \langle x|p\rangle^* = A^* e^{-ipx/\hbar}. \tag{8.82}$$

Now we perform a basis transformation of the state $|\psi\rangle$ in (8.66) by operating on it with the identity operator in momentum space

$$|\psi\rangle = I|\psi\rangle = \left(\int_{-\infty}^{\infty} dp\,|p\rangle\langle p|\right)\left(\int_{-\infty}^{\infty} dx\,\psi(x)\,|x\rangle\right) \tag{8.83}$$

$$= \int_{-\infty}^{\infty} dp \left(\int_{-\infty}^{\infty} dx\,\psi(x)\,\langle p|x\rangle\right)|p\rangle \tag{8.84}$$

$$= \int_{-\infty}^{\infty} dp \left(A^* \int_{-\infty}^{\infty} dx\,e^{-ipx/\hbar}\,\psi(x)\right)|p\rangle. \tag{8.85}$$

Comparing the last line with (8.68), we get

$$\tilde{\psi}(p) = A^* \int_{-\infty}^{\infty} dx\,e^{-ipx/\hbar}\,\psi(x). \tag{8.86}$$

A similar analysis can be used to find the inverse transform

$$\psi(x) = A \int_{-\infty}^{\infty} dp\,e^{ipx/\hbar}\,\tilde{\psi}(p). \tag{8.87}$$

The constant A can be found by using the fact that if we transform a wave function using (8.87) followed by the inverse transform (8.86), we should return to the original wave function. This gives $|A|^2 = 1/(2\pi\hbar)$ and so the transform and its inverse can be written as

$$\tilde{\psi}(p) = \frac{1}{\sqrt{2\pi\hbar}} \int_{-\infty}^{\infty} dx\,e^{-ipx/\hbar}\,\psi(x) \tag{8.88}$$

$$\psi(x) = \frac{1}{\sqrt{2\pi\hbar}} \int_{-\infty}^{\infty} dp\,e^{ipx/\hbar}\,\tilde{\psi}(p). \tag{8.89}$$

Apart from a factor, these are just the transforms (8.75) and (8.77), which means we can convert between the position and momentum representations of the wave function of a system simply by taking the Fourier transform and its inverse. One consequence of this is that a particle that is well-localised in space will have a broad momentum distribution and vice versa. We will study this effect more formally when we revisit the Heisenberg uncertainty relation later.

Let us now introduce the inner product of two state vectors $|\psi\rangle$ and $|\phi\rangle$ in position space. In analogy with the inner product of two states of finite dimension, we have

$$\langle\phi|\psi\rangle = \left(\int_{-\infty}^{\infty} dx\, \phi^*(x)\langle x|\right)\left(\int_{-\infty}^{\infty} dx'\, \psi(x')|x'\rangle\right) \quad (8.90)$$

$$= \int_{-\infty}^{\infty} dx \int_{-\infty}^{\infty} dx' \phi^*(x)\psi(x')\langle x|x'\rangle \quad (8.91)$$

$$= \int_{-\infty}^{\infty} dx \int_{-\infty}^{\infty} dx' \phi^*(x)\psi(x')\delta(x-x') \quad (8.92)$$

$$= \int_{-\infty}^{\infty} dx\, \phi^*(x)\psi(x). \quad (8.93)$$

Note that this is the same expression for the overlap between two states that we introduced in Section 3.7. We can also demonstrate that $\psi(x) = \langle x|\psi\rangle$ (as noted earlier) by direct calculation,

$$\langle x|\psi\rangle = \langle x|\left(\int_{-\infty}^{\infty} dx'\, \psi(x')|x'\rangle\right) \quad (8.94)$$

$$= \int_{-\infty}^{\infty} dx'\, \psi(x')\langle x|x'\rangle \quad (8.95)$$

$$= \int_{-\infty}^{\infty} dx'\, \psi(x')\delta(x-x') = \psi(x). \quad (8.96)$$

Using (8.93), a state is normalised when

$$\langle\psi|\psi\rangle = \int_{-\infty}^{\infty} dx\, |\psi(x)|^2 = 1. \quad (8.97)$$

We also know from the third postulate of quantum mechanics that the probability density of finding a particle at position x_0, given that it was prepared in state $|\psi\rangle$, is

$$P_\psi(x_0) = |\langle x_0|\psi\rangle|^2 = |\psi(x_0)|^2. \quad (8.98)$$

Similarly, the probability density of measuring a particle with momentum p_0 is

$$P_\psi(p_0) = |\langle p_0|\psi\rangle|^2 = |\tilde{\psi}(p_0)|^2. \quad (8.99)$$

8.9. Position and Momentum Operators Revisited

In Section 3.7 we introduced expressions for the operators for position, momentum and energy in the position space representation. In this section we will justify their forms. To do so it is helpful to calculate the expectation values of these operators. Let us start with position. The expectation value is

$$\langle x\rangle = \langle\psi|\hat{x}|\psi\rangle = \langle\psi|\left(\int_{-\infty}^{\infty} dx\, x|x\rangle\langle x|\right)|\psi\rangle \tag{8.100}$$

$$= \int_{-\infty}^{\infty} dx\, \langle\psi|x\rangle x\langle x|\psi\rangle \tag{8.101}$$

$$= \int_{-\infty}^{\infty} dx\, \psi^*(x)\, x\, \psi(x), \tag{8.102}$$

which is what we saw in Section 3.7. Now, we have seen from Eq. (8.44) that the expression for the expectation value for some observable A with corresponding operator $\hat{A}$ is $\langle A\rangle = \langle\psi|\hat{A}|\psi\rangle$, which is written is position representation as

$$\langle A\rangle = \int_{-\infty}^{\infty} \psi^*(x)\, \hat{A}\, \psi(x)\, dx. \tag{8.103}$$

Comparing (8.102) with (8.103) we see that the position operator $\hat{x}$, when expressed in the position representation, is simply x.

Let us now carry out a similar calculation to determine the momentum operator in position space. Calculating the expectation of momentum for a general state $|\psi\rangle$, we get

$$\begin{aligned}
\langle\psi|\hat{p}|\psi\rangle &= \left(\int_{-\infty}^{\infty} dx'\psi^*(x')\langle x'|\right)\left(\int_{-\infty}^{\infty} dp\, p|p\rangle\langle p|\right)\left(\int_{-\infty}^{\infty} dx\psi(x)|x\rangle\right)\\
&= \int_{-\infty}^{\infty} dx' \int_{-\infty}^{\infty} dp \int_{-\infty}^{\infty} dx\, \psi^*(x')\, p\, \psi(x)\langle x'|p\rangle\langle p|x\rangle\\
&= \frac{1}{2\pi\hbar}\int_{-\infty}^{\infty} dx'\psi^*(x')\int_{-\infty}^{\infty} dp\left(\int_{-\infty}^{\infty} dx\, p\, \psi(x)e^{-ip(x-x')/\hbar}\right).
\end{aligned}$$

Evaluating the integral in round brackets using integration by parts and making use of the integral identity,

$$\int_{-\infty}^{\infty} dp\, e^{-ip(x-x')/\hbar} = 2\pi\hbar\delta(x-x'), \tag{8.104}$$

the result after a few lines of working (try it) is

$$\langle p\rangle = \int_{-\infty}^{\infty} dx\, \psi^*(x)\left(-i\hbar\frac{d}{dx}\right)\psi(x). \tag{8.105}$$

This is the same as the expression that we saw in Section 3.7. Comparing this expression with (8.103), we see that the momentum operator in position space can be written as $-i\hbar\frac{d}{dx}$. Finally, since the Hamiltonian is a function of $\hat{p}$ and $\hat{x}$, we can use the correspondence principle to write down an expression for the

Hamiltonian in the position representation. This involves simply replacing each operator by its representation in position space. Summarising, we have

$$\hat{x} \longrightarrow x \tag{8.106}$$

$$\hat{p} \longrightarrow -i\hbar\frac{d}{dx} \tag{8.107}$$

$$\hat{H} = \frac{\hat{p}^2}{2M} + V(\hat{x}) \longrightarrow -\frac{\hbar^2}{2M}\frac{d^2}{dx^2} + V(x). \tag{8.108}$$

8.10. The Schrödinger Equation Revisited

Back in Section 3.1, we saw that the time-dependent one-dimensional Schrödinger equation can be written in the position representation as

$$-\frac{\hbar^2}{2M}\frac{\partial^2\Psi(x,t)}{\partial x^2} + V(x)\Psi(x,t) = i\hbar\frac{\partial\Psi(x,t)}{\partial t}. \tag{8.109}$$

This describes the unitary time evolution of a particle of mass M in a potential $V(x)$ (see the fourth postulate of quantum mechanics). From the result in (8.108), we see that this can be written in a very compact form using Dirac notation,

$$i\hbar\frac{\partial}{\partial t}|\psi\rangle = \hat{H}|\psi\rangle, \tag{8.110}$$

where

$$\hat{H} = -\frac{\hbar^2}{2M}\frac{d^2}{dx^2} + V(x). \tag{8.111}$$

This means that an equivalent expression for the Hamiltonian (or total energy) operator is

$$\hat{H} \longrightarrow i\hbar\frac{\partial}{\partial t}, \tag{8.112}$$

as introduced in Section 3.7. The Schrödinger equation in Dirac notation is useful not only because it is compact, but also because it is a general form that holds independently of the representation of the state. By comparison, Eq. (8.109) is the Schrödinger equation only for the position representation.

Similarly, the Dirac form of the time-independent Schrödinger equation is given by

$$\hat{H}|\psi\rangle = E|\psi\rangle. \tag{8.113}$$

This is an eigenvalue equation and $\{E\}$ is the set of energies corresponding to the eigenstates. As we have seen earlier, these eigenstates are stationary solutions of the Schrödinger equation, i.e. they do not evolve with time. We have studied the solution to this equation in a number of physical contexts (e.g. the infinite square well, the quantum harmonic oscillator and the hydrogen atom) to find the allowed energies and states.

The connection between stationary solutions and time-dependent solutions of the Dirac form of the Schrödinger equation can be seen as follows. Suppose the state of a system at time $t = 0$ is

$$|\psi(0)\rangle = \sum_{i=0}^{n-1} \alpha_i |\psi_i\rangle, \tag{8.114}$$

where each state $|\psi_i\rangle$ is a solution of the time-independent Schrödinger equation with corresponding energy E_i, i.e. for every i,

$$\hat{H}|\psi_i\rangle = E_i|\psi_i\rangle. \tag{8.115}$$

Then the solution of the time-dependent Schrödinger equation (8.110) is simply

$$|\psi(t)\rangle = \sum_{i=0}^{n-1} \alpha_i \, e^{-iE_i t/\hbar} \, |\psi_i\rangle. \tag{8.116}$$

This can be checked by substitution into (8.110).

8.11. The Uncertainty Principle Revisited

Heisenberg's uncertainty principle is a statement about how accurately two different observables can be simultaneously measured for a quantum system. For certain pairs of variables, there is a fundamental limit to the product of the precisions that can be attained. One such pair of observables is position and momentum. Here we want to use our operator formalism to revisit the uncertainty principle and rewrite it in terms of general observables.

Suppose we have an ensemble of systems all prepared in the same state $|\psi\rangle$ and that we wish to measure two different observables A and B on this ensemble. The precision of these measurements is given by the standard deviation of the outcomes, i.e.

$$\Delta A = \sqrt{\langle\psi|\hat{A}^2|\psi\rangle - \langle\psi|\hat{A}|\psi\rangle^2} \tag{8.117}$$

$$\Delta B = \sqrt{\langle\psi|\hat{B}^2|\psi\rangle - \langle\psi|\hat{B}|\psi\rangle^2}. \tag{8.118}$$

Heisenberg's uncertainty principle says that the product of these measurement precisions is bounded by,

$$\Delta A \, \Delta B \geq \frac{1}{2}\left|\langle\psi|(\hat{A}\hat{B} - \hat{B}\hat{A})|\psi\rangle\right|. \tag{8.119}$$

The right hand side of this inequality need not be zero, as we can see by taking the example of position and momentum. To calculate this, it is convenient to transform into the position representation of the operators

$$\begin{aligned}
\langle\psi|(\hat{x}\hat{p} - \hat{p}\hat{x})|\psi\rangle &= \int_{-\infty}^{\infty} dx\, \psi^*(x) \left[-i\hbar x \frac{d}{dx} + i\hbar \frac{d}{dx} x\right] \psi(x) \\
&= i\hbar \int_{-\infty}^{\infty} dx\, \psi^*(x) \left[-x \frac{d}{dx}\psi(x) + \frac{d}{dx}(x\psi(x))\right] \\
&= i\hbar \int_{-\infty}^{\infty} dx\, \psi^*(x) \left[-x \frac{d}{dx}\psi(x) + x \frac{d}{dx}\psi(x) + \psi(x)\right] \\
&= i\hbar \int_{-\infty}^{\infty} dx\, \psi^*(x)\, \psi(x) \\
&= i\hbar.
\end{aligned} \tag{8.120}$$

Substituting this result into (8.119) we get

$$\Delta x \Delta p \geq \frac{\hbar}{2}, \tag{8.121}$$

which is the well-known form of the uncertainty relation for momentum and position. This result implies that, when the position of a quantum mechanical particle can be measured with a very high precision, then the momentum of the particle is spread over a wide range of values. Conversely, a particle with a well defined momentum cannot be well localised. This formalises the observation that we made when considering Fourier transforms in Section 8.8.

8.12. Pure and Mixed States

Up until now, we have only discussed states $|\psi\rangle$ that are known exactly. These are called pure states and can be written as a superposition of vectors in some basis space. Another way of representing them is in terms of a density matrix, which is given by $\rho = |\psi\rangle\langle\psi|$. More generally, if a state cannot be written in this form it is called a mixed state. In this case, it can be written as a classical mixture of different pure states.

We can think of this distinction in terms of the difference between quantum and classical physics. Suppose we had a paper bag filled with 50 red marbles and 50 blue ones. Without opening the bag, we can describe its (classical) state, i.e. our total knowledge of the system, as an equal mixture of the pure states representing red and blue marbles. In density matrix notation, this can be written as

$$\rho = \frac{1}{2}\left(|\text{red}\rangle\langle\text{red}| + |\text{blue}\rangle\langle\text{blue}|\right) \tag{8.122}$$

or, as a matrix in the basis $\{|\text{red}\rangle, |\text{blue}\rangle\}$, as

$$\rho = \frac{1}{2}\begin{pmatrix} 1 & 0 \\ 0 & 1 \end{pmatrix}. \tag{8.123}$$

By contrast, suppose we had a single electron that we know is in an equal superposition of spin up and spin down. The density matrix for this (pure) state can be written as

$$\rho = \frac{1}{2}\left(|\uparrow\rangle\langle\uparrow| + |\downarrow\rangle\langle\downarrow| + |\uparrow\rangle\langle\downarrow| + |\downarrow\rangle\langle\uparrow|\right) \tag{8.124}$$

or, in the basis $\{|\uparrow\rangle, |\downarrow\rangle\}$, as

$$\rho = \frac{1}{2}\begin{pmatrix} 1 & 1 \\ 1 & 1 \end{pmatrix}. \tag{8.125}$$

Density matrices are a very useful tool for studying real-world situations. They allow us to extend the range of applicability of quantum mechanics to include systems for which we don't know the exact state, i.e. when the system is in some classical distribution of pure states. In general, if a system is prepared in the pure state $\rho_i = |\psi_i\rangle\langle\psi_i|$ with probability p_i, then the density matrix of the state is simply given by

$$\rho = \sum_i p_i \rho_i. \tag{8.126}$$

Density matrices have two important properties. The first is that their trace, i.e. the sum of the diagonal elements, is always one. This really is just a statement of the normalisation condition. In other words, the system must be found in one of the states available to it. You can easily see that the trace is one for the cases (8.123) and (8.125) we considered above. The second important property is that density matrices always have positive eigenvalues. We can see this from (8.126) – the total density matrix is diagonal if written in the basis $\{|\psi_i\rangle\}$, and the corresponding eigenvalues are $\{p_i\}$, which are all positive, because they are just classical probabilities.

Given a density matrix ρ, it is a relatively straightforward exercise to establish whether it corresponds to a pure or a mixed state. For pure states the trace of the density matrix squared is always one, i.e. $\mathrm{Tr}\{\rho^2\} = 1$. We can see this as follows. The density matrix for a pure state $|\psi\rangle$ is $\rho = |\psi\rangle\langle\psi|$. Squaring this we get

$$\rho^2 = |\psi\rangle\langle\psi|\psi\rangle\langle\psi| = |\psi\rangle\langle\psi| = \rho, \tag{8.127}$$

where we have made use of the normalisation $\langle\psi|\psi\rangle = 1$. So, for a pure state, $\rho^2 = \rho$, which means that $\mathrm{Tr}\{\rho^2\} = \mathrm{Tr}\{\rho\}$ and, since we know that the trace of any density matrix is always one, it follows that $\mathrm{Tr}\{\rho^2\} = 1$. For a mixed state, however, we have $\mathrm{Tr}\{\rho^2\} < 1$. It is easy to check these results for the cases discussed above. For the mixed state (8.123), we have $\mathrm{Tr}\{\rho^2\} = 1/4 < 1$ and for the pure state (8.125), we have $\mathrm{Tr}\{\rho^2\} = 1$, as expected.

There are a number of subtleties associated with density matrices. For example, there are infinitely many ways of writing down one and the same density matrix, but all of them are identical as far as measurement outcomes on the system are concerned. Suppose, for example we had an electron in an equal mixture of $|\uparrow\rangle$ and $|\downarrow\rangle$, i.e.

$$\rho = \frac{1}{2}\left(|\uparrow\rangle\langle\uparrow| + |\downarrow\rangle\langle\downarrow|\right). \tag{8.128}$$

This can equally well be written as[2]

$$\rho = \frac{1}{2}\left(|+\rangle\langle+| + |-\rangle\langle-|\right), \tag{8.129}$$

where $|\pm\rangle = (|\uparrow\rangle \pm |\downarrow\rangle)/\sqrt{2}$. There is no experimental procedure that can distinguish between these two cases since they both represent exactly the same physical system! This notion takes a bit of getting used to.

For now, we will go back to thinking only about pure states, but we will meet mixed states again in the final chapter of the book when we come to discuss an important feature of quantum mechanics called entanglement.

8.13. Annihilation and Creation Operators

One operator that is particularly useful and widely used in quantum physics is the annihilation operator, a. Its adjoint, $a^\dagger$, is called the creation operator. As their names suggest, these operators create or annihilate quanta from a system. They are the same as the raising and lowering operators we introduced in Section 4.6.

We can understand where the annihilation and creation operators come from by revisiting the quantum harmonic oscillator discussed in Section 4.2. Recall that the Hamiltonian for this system is

$$H = \frac{\hat{p}^2}{2M} + \frac{1}{2}M\omega^2\hat{x}^2. \tag{8.130}$$

If we then make the following substitutions for the position and momentum operators[3]

$$\hat{x} = \sqrt{\frac{\hbar}{2M\omega}}\left(a + a^\dagger\right) \tag{8.131}$$

$$\hat{p} = i\sqrt{\frac{\hbar M\omega}{2}}\left(a^\dagger - a\right), \tag{8.132}$$

we obtain the much simpler form for the Hamiltonian,

$$H = \frac{1}{2}(a^\dagger a + aa^\dagger)\hbar\omega. \tag{8.133}$$

[2]We saw a very similar idea in the context of identity matrices in Section 8.3.
[3]We will see why shortly.

We can find the commutation relation for a and $a^\dagger$ by calculating $[\hat{x}, \hat{p}]$ using (8.131) and (8.132),

$$\begin{aligned}[\hat{x}, \hat{p}] &= \frac{i\hbar}{2}\left[a + a^\dagger, a^\dagger - a\right] \\ &= \frac{i\hbar}{2}\left(\left[a, a^\dagger\right] - \left[a^\dagger, a\right]\right) = i\hbar\left[a, a^\dagger\right].\end{aligned} \tag{8.134}$$

Now since we know that $[\hat{x}, \hat{p}] = i\hbar$, we must have

$$\left[a, a^\dagger\right] = 1. \tag{8.135}$$

Now we can use this result to rewrite (8.133) as

$$H = \left(a^\dagger a + 1/2\right)\hbar\omega. \tag{8.136}$$

Comparing this with the form of the energy for the nth level in a quantum harmonic oscillator, $E_n = (n+1/2)\hbar\omega$, we see that $a^\dagger a$ is simply the operator for the energy level. Since each energy level is equally spaced, $a^\dagger a$ can equivalently be thought of as the operator for the number of energy quanta, $\hbar\omega$. For this reason, it is often called the number operator.

Let us now think about the effect of the annihilation and creation operators. We take the example of the creation operator acting on the ground state of the quantum harmonic oscillator. From (8.131) and (8.132) we can write the creation operator as

$$\begin{aligned}a^\dagger &= \sqrt{\frac{M\omega}{2\hbar}}\left(\hat{x} - \frac{i}{M\omega}\hat{p}\right) \\ &= \frac{1}{\sqrt{2}b}\left(\hat{x} - b^2\frac{d}{dx}\right),\end{aligned} \tag{8.137}$$

where $b = \sqrt{\hbar/(M\omega)}$ and we have used $\hat{p} = -i\hbar d/dx$. The ground state of the harmonic oscillator is (see Section 4.2)

$$\psi_0(x) = \left(\frac{1}{\sqrt{\pi}b}\right)^{1/2} e^{-x^2/2b^2}, \tag{8.138}$$

so

$$\begin{aligned}a^\dagger\psi_0(x) &= \left(\frac{1}{2\sqrt{\pi}b^3}\right)^{1/2}\left(\hat{x} - b^2\frac{d}{dx}\right)e^{-x^2/2b^2} \\ &= \left(\frac{2}{\sqrt{\pi}b^3}\right)^{1/2} xe^{-x^2/2b^2} = \psi_1(x),\end{aligned} \tag{8.139}$$

where the final step is given by (4.15). So we see that the creation operator has taken us from the ground state ($n = 0$) to the first excited state ($n = 1$) or, equivalently, it has *created* one additional quantum in the system. A similar calculation shows that the annihilation operator reduces the number of quanta by one.

This relationship is conveniently generalised by rewriting the system in the basis of the energy level or, equivalently, the number of quanta or particles present, i.e. the ground state is $|0\rangle$ and the Nth excited state is $|N\rangle$. This is a commonly-used basis called the number or Fock[4] basis. In it, the action of the annihilation and creation operators are given by

$$a|N\rangle = \sqrt{N}|N-1\rangle \tag{8.140}$$

$$a^\dagger|N\rangle = \sqrt{N+1}|N+1\rangle. \tag{8.141}$$

From this formalism it is easy to see that $a^\dagger a$ is the number operator, as discussed above, by considering the operation of $a^\dagger a$ on the Fock state with N particles

$$\begin{aligned} a^\dagger a|N\rangle &= a^\dagger\sqrt{N}|N-1\rangle = \sqrt{N}(a^\dagger|N-1\rangle) \\ &= \sqrt{N}(\sqrt{N}|N\rangle) = N|N\rangle. \end{aligned} \tag{8.142}$$

From this we see that the Fock state with N particles is an eigenstate of the operator $a^\dagger a$ with eigenvalue N, i.e. $a^\dagger a$ is the number operator, as expected.

The annihilation and creation operators are very useful and widely used in many different contexts within quantum physics. We will use them in the next section to study the Mach–Zehnder interferometer and will meet them again in Section 10.5 when we consider quantum field theory.

8.14. The Mach–Zehnder Interferometer

Let us finish the chapter by revisiting the example of the Mach–Zehnder interferometer that we discussed way back in Chapter 1. We are now able to apply some of the techniques we have learned from quantum mechanics to analyse it in a much more precise and formal way.

It is helpful to start by considering the action of a 50:50 beam splitter in Dirac notation (see Fig. 8.1). In the Fock state basis, a photon initially on path 1 becomes

$$\frac{1}{\sqrt{2}}\left(|1\rangle_2|0\rangle_3 + i|0\rangle_2|1\rangle_3\right), \tag{8.143}$$

after passing through the beam splitter. The first term in (8.143) represents one photon on path 2 and none on path 3; the second term is the converse. The factor of i is due to the $\pi/2$ phase change that the photon experiences on reflection[5]. Instead of writing this as a transformation of the state, we can

[4] Named after Vladimir Fock (1898–1974), a prominent Soviet physicist who made many significant contributions not only to quantum physics, but also to geophysical exploration.

[5] We neglected this phase shift in our treatment in Chapter 1 to avoid unnecessary complications, but now we're doing things properly.

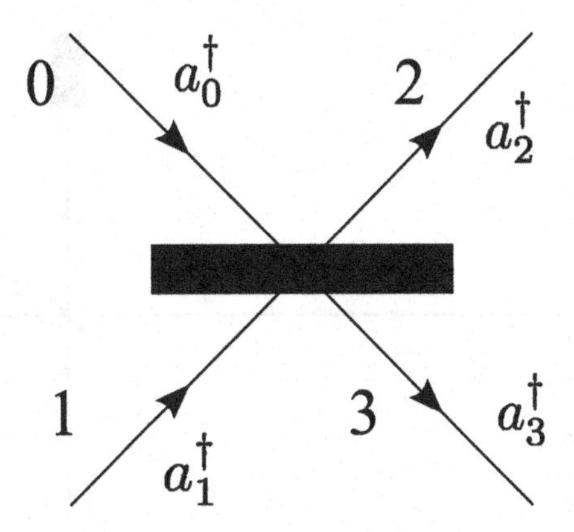

Fig. 8.1. A 50:50 beam splitter with input modes 0 and 1 and output modes 2 and 3. The creation operators for a particle in each of these modes are respectively $a_0^\dagger$, $a_1^\dagger$, $a_2^\dagger$, and $a_3^\dagger$.

equally well write it in terms of operators. We can see this as follows. The initial state can be written as

$$|0\rangle_0|1\rangle_1 = a_1^\dagger|0\rangle|0\rangle \tag{8.144}$$

and the final state as

$$\frac{1}{\sqrt{2}}\left(|1\rangle_2|0\rangle_3 + i|0\rangle_2|1\rangle_3\right) = \frac{1}{\sqrt{2}}\left(a_2^\dagger + ia_3^\dagger\right)|0\rangle|0\rangle, \tag{8.145}$$

where $a_j^\dagger$ is the creation operator for a photon in mode $j \in \{0, 1, 2, 3\}$. From (8.144) and (8.145), we see that the transformation of the state by the 50:50 beam splitter is given by

$$a_1^\dagger|0\rangle|0\rangle \longrightarrow \frac{1}{\sqrt{2}}\left(a_2^\dagger + ia_3^\dagger\right)|0\rangle|0\rangle, \tag{8.146}$$

i.e.

$$a_1^\dagger \longrightarrow \frac{1}{\sqrt{2}}\left(a_2^\dagger + ia_3^\dagger\right). \tag{8.147}$$

Similarly, if we start with a photon initially in mode 0, we can obtain the operator transform

$$a_0^\dagger \longrightarrow \frac{1}{\sqrt{2}}\left(ia_2^\dagger + a_3^\dagger\right). \tag{8.148}$$

These transformations are very useful and allow us to calculate how very general quantum states are transformed by beam splitters. To do this, we rewrite the state in terms of creation operators, then transform the operators according to the rules above.

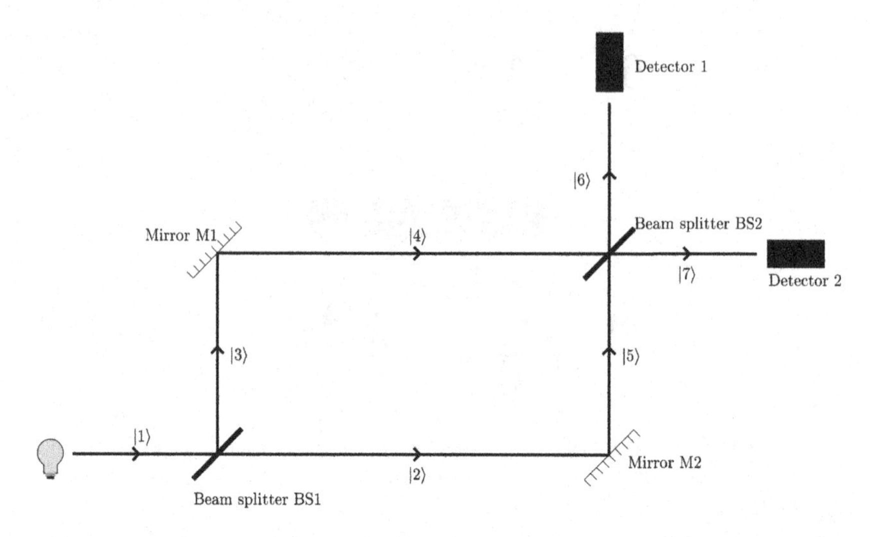

Fig. 8.2. Mach–Zehnder Interferometer. The figure is reproduced here from Chapter 1.

As an example, consider the case of a 50:50 beam splitter with a single photon incident on each input. This state can be written as

$$|1\rangle_0|1\rangle_1 = a_0^\dagger a_1^\dagger|0\rangle|0\rangle. \tag{8.149}$$

After passing through the beam splitter, the operators are transformed according to the rules (8.147) and (8.148), which gives

$$\frac{1}{2}\left(ia_2^\dagger + a_3^\dagger\right)\left(a_2^\dagger + ia_3^\dagger\right)|0\rangle|0\rangle = \frac{i}{2}\left(a_2^{\dagger 2} + a_3^{\dagger 2}\right)|0\rangle|0\rangle. \tag{8.150}$$

The final state can be found by operating on the vacuum states with the creation operators and the full transformation can, therefore, be summarised as

$$|1\rangle_0|1\rangle_1 \longrightarrow \frac{i}{\sqrt{2}}\left(|2\rangle_2|0\rangle_3 + |0\rangle_2|2\rangle_3\right). \tag{8.151}$$

We see that the final state is a superposition of both photons exiting at one output and both exiting at the other. Strangely, we never get one photon in each output port. This is called the Hong–Ou–Mandel effect and is further evidence of the tendency for bosons to want to clump together[6] that we saw earlier in Chapter 6.

[6]By complete contrast, if we had used fermions rather than photons (which are bosons), we would have always seen one particle in each output port. The reason for this is the Pauli exclusion principle, which prevents two particles from being in the same mode. This simple set-up provides a clear and dramatic example of the different behaviour of bosons and fermions.

Now that we understand the operation of a beam splitter, we can go on to consider the Mach–Zehnder interferometer. For convenience, we have reproduced the Mach–Zehnder interferometer from Chapter 1 here as Fig. 8.2. Suppose that we start with a single photon in input mode 1, i.e. the initial state is $|0\rangle_0|1\rangle_1 = a_1^\dagger|0\rangle|0\rangle$. Let us consider how the operator is changed at each step of the interferometer. We have seen that the first beam splitter (BS1) makes the transform given by (8.147). Next, modes 2 and 3 meet mirrors M2 and M1 respectively, these make the transforms:

$$a_2^\dagger \xrightarrow{M2} ia_5^\dagger \tag{8.152}$$
$$a_3^\dagger \xrightarrow{M1} ia_4^\dagger, \tag{8.153}$$

where the factor of i is just the $\pi/2$ phase shift due to reflection that we saw earlier. Finally, modes 4 and 5 encounter the second beam splitter (BS2) and, from our analysis above, we know this operation makes the transformations:

$$a_4^\dagger \xrightarrow{BS2} \frac{1}{\sqrt{2}}\left(ia_6^\dagger + a_7^\dagger\right) \tag{8.154}$$
$$a_5^\dagger \xrightarrow{BS2} \frac{1}{\sqrt{2}}\left(a_6^\dagger + ia_7^\dagger\right). \tag{8.155}$$

Combining all these steps, we get the transformation of the complete Mach–Zehnder interferometer:

$$a_1^\dagger \xrightarrow{BS1} \frac{1}{\sqrt{2}}\left(a_2^\dagger + ia_3^\dagger\right) \tag{8.156}$$
$$\xrightarrow{M1,M2} \frac{1}{\sqrt{2}}\left(ia_5^\dagger - a_4^\dagger\right) \tag{8.157}$$
$$\xrightarrow{BS2} \frac{1}{2}\left(i[a_6^\dagger + ia_7^\dagger] - [ia_6^\dagger + a_7^\dagger]\right) = -a_7^\dagger. \tag{8.158}$$

In other words, if we start with a single particle in mode 1, the output of the interferometer is always a single particle in mode 7,

$$|0\rangle_0|1\rangle_1 \longrightarrow |0\rangle_6|1\rangle_7. \tag{8.159}$$

This is a more formal way of arriving at the surprising conclusion that we discussed in the introduction to the book.

It is also worthwhile considering a more general analysis of the interferometer. We have considered the case where the two paths that a particle can take between the beam splitters have the same length. Suppose instead that one path is slightly longer than the other. This means that the particle will have a different phase when it arrives at BS2 depending on which path it took . We can easily modify our analysis to account for this, by including an extra phase, θ, on (say) mode 4. The transformation of $a_1^\dagger$ then becomes

$$a_1^\dagger \xrightarrow{BS1} \frac{1}{\sqrt{2}}\left(a_2^\dagger + ia_3^\dagger\right) \tag{8.160}$$

$$\stackrel{M1,M2,\theta}{\longrightarrow} \frac{1}{\sqrt{2}}\left(ia_5^\dagger - e^{i\theta}\, a_4^\dagger\right) \tag{8.161}$$

$$\stackrel{BS2}{\longrightarrow} \frac{1}{2}\left(i[a_6^\dagger + ia_7^\dagger] - e^{i\theta}\left[ia_6^\dagger + a_7^\dagger\right]\right) \tag{8.162}$$

$$= \sin\left(\frac{\theta}{2}\right) a_6^\dagger - \cos\left(\frac{\theta}{2}\right) a_7^\dagger, \tag{8.163}$$

where we have neglected an unimportant overall phase.

So, if we start with a single particle in mode 1, at the output we would expect to find it in mode 6 with probability $P_6 = \sin^2(\theta/2)$ and in mode 7 with probability $P_7 = \cos^2(\theta/2)$. This means that by sending a whole stream of particles through the interferometer one after the other and detecting the number at each output we can infer the value of θ. Since θ varies by 2π for a path-length difference corresponding to one wavelength of the light, this technique enables us to measure relative distances to within a fraction of the wavelength of light. This amazing precision is why interferometers have been such an important tool in the field of metrology.

8.15. Perturbation Theory

We round off this chapter by introducing a very useful tool called perturbation theory that enables us tackle difficult problems in quantum physics. Perturbation theory is important because, just like classical physics, there are few physically interesting problems that can be solved exactly. We, therefore, need to develop an armoury of tools for finding approximate solutions. Perturbation theory is just such a tool and allows us to find approximate eigenstates and eigenvalues to problems that are 'close' to other problems that we already know the solution to. Here, we will consider applying perturbation theory only to time-independent systems and will consider only the first order corrections.

To understand perturbation theory[7], let us suppose that the time-independent Hamiltonian of a system can be expressed as

$$H = H_0 + \lambda H', \tag{8.164}$$

where the solutions to the Schrödinger equation corresponding to the unperturbed Hamiltonian, H_0,

$$H_0|\psi_i^{(0)}\rangle = E_i^{(0)}|\psi_i^{(0)}\rangle, \tag{8.165}$$

are known and $\lambda H'$ is a small perturbation. We assume that the solutions to the unperturbed equation form a complete set of orthogonal and normalised states. The quantity λ is a real number that is used to keep track of the various orders of the perturbation calculation, as we shall see. Setting $\lambda = 0$ returns

[7] Strictly, we should say quantum perturbation theory since perturbation theory has a long history in classical mechanics, where it first distinguished itself as a great tool for calculating the orbits of planets.

the unperturbed result and setting $\lambda = 1$ gives the result for the perturbed Hamiltonian we are interested in, $H_0 + H'$. The eigenvalue problem we want to solve is

$$H|\psi_i\rangle = E_i|\psi_i\rangle. \tag{8.166}$$

The basic idea of perturbation theory is to assume that both the eigenvalues and the eigenstates of H can be expanded in powers of the perturbation parameter, λ, i.e.

$$E_i = \sum_{j=0}^{\infty} \lambda^j E_i^{(j)}, \tag{8.167}$$

$$|\psi_i\rangle = \sum_{j=0}^{\infty} \lambda^j |\psi_i^{(j)}\rangle. \tag{8.168}$$

Substituting these expressions into the eigenvalue equation (8.166), we obtain,

$$(H_0 + \lambda H')(|\psi_i^{(0)}\rangle + \lambda|\psi_i^{(1)}\rangle + \cdots) = (E_i^{(0)} + \lambda E_i^{(1)} + \cdots) \times (|\psi_i^{(0)}\rangle + \lambda|\psi_i^{(1)}\rangle + \cdots). \tag{8.169}$$

We now equate the coefficients of equal powers of λ on both sides of the equation. For λ^0, as expected, we get

$$H_0|\psi_i^{(0)}\rangle = E_i^{(0)}|\psi_i^{(0)}\rangle. \tag{8.170}$$

For λ^1, we get

$$H_0|\psi_i^{(1)}\rangle + H'|\psi_i^{(0)}\rangle = E_i^{(0)}|\psi_i^{(1)}\rangle + E_i^{(1)}|\psi_i^{(0)}\rangle. \tag{8.171}$$

We could, of course, carry on for higher orders of λ, but will not do so here.

In order to calculate the first-order energy correction, $E_i^{(1)}$, we premultiply equation (8.171) by $\langle\psi_i^{(0)}|$. This gives

$$\langle\psi_i^{(0)}|H_0 - E_i^{(0)}|\psi_i^{(1)}\rangle + \langle\psi_i^{(0)}|H' - E_i^{(1)}|\psi_i^{(0)}\rangle = 0. \tag{8.172}$$

Now, we make use of the fact that H_0 is Hermitian, this means that $H_0 = H_0^\dagger$. So,

$$\langle\psi_i^{(0)}|H_0|\psi_i^{(1)}\rangle = \langle\psi_i^{(0)}|H_0^\dagger|\psi_i^{(1)}\rangle = (\langle\psi_i^{(1)}|H_0|\psi_i^{(0)}\rangle)^\dagger \tag{8.173}$$

$$= (E_i^{(0)}\langle\psi_i^{(1)}|\psi_i^{(0)}\rangle)^\dagger = E_i^{(0)}\langle\psi_i^{(0)}|\psi_i^{(1)}\rangle, \tag{8.174}$$

where we have made use of (8.170) and the fact that $E_i^{(0)}$ is real. Substituting this result into (8.172), we see that the first term vanishes and we are left with

$$\langle\psi_i^{(0)}|E_i^{(1)}|\psi_i^{(0)}\rangle = \langle\psi_i^{(0)}|H'|\psi_i^{(0)}\rangle, \tag{8.175}$$

i.e.

$$E_i^{(1)} = \langle\psi_i^{(0)}|H'|\psi_i^{(0)}\rangle. \tag{8.176}$$

This is a very important result and says that the first-order correction to the energy is just the expectation value of the perturbation H' over an ensemble of the corresponding unperturbed states. Higher order corrections can also be obtained in a similar way.

We would now like to turn our attention to how the states are changed due to the presence of the perturbation. Since the unperturbed eigenstates form a complete basis, we know that we can expand the first-order correction to the eigenstate in terms of the unperturbed states, i.e.

$$|\psi_i^{(1)}\rangle = \sum_k a_{ik}^{(1)}\,|\psi_k^{(0)}\rangle. \tag{8.177}$$

Substituting into (8.171), we obtain

$$(H_0 - E_i^{(0)})\sum_k a_{ik}^{(1)}\,|\psi_k^{(0)}\rangle + (H' - E_i^{(1)})|\psi_i^{(0)}\rangle = 0. \tag{8.178}$$

Premultiplying by $\langle\psi_l^{(0)}|$, we get

$$(E_l^{(0)} - E_i^{(0)})a_{il}^{(1)} + \langle\psi_l^{(0)}|H'|\psi_i^{(0)}\rangle - E_i^{(1)}\delta_{il} = 0. \tag{8.179}$$

For $i = l$ this equation reduces to $E_i^{(1)} = \langle\psi_i^{(0)}|H'|\psi_i^{(0)}\rangle$ just as we saw earlier in equation (8.176). For $i \neq l$,

$$a_{il}^{(1)} = \frac{\langle\psi_l^{(0)}|H'|\psi_i^{(0)}\rangle}{E_i^{(0)} - E_l^{(0)}}. \tag{8.180}$$

It turns out that without loss of generality we can set all the diagonal terms to zero[8], i.e. $a_{ii}^{(1)} = 0$. Substituting these values into (8.177), we obtain an expression for the first-order correction to the eigenstates

$$|\psi_i^{(1)}\rangle = \sum_{k\neq i} \frac{\langle\psi_k^{(0)}|H'|\psi_i^{(0)}\rangle}{E_i^{(0)} - E_k^{(0)}}|\psi_k^{(0)}\rangle. \tag{8.181}$$

This along with (8.176) are the two key results in first-order time-independent perturbation theory.

Perhaps this will all become clearer by considering an example. Suppose we have a particle of mass M trapped in a potential of the form

$$V(x) = \frac{1}{2}M\omega^2 x^2 + \lambda \exp\left(-\frac{M\omega x^2}{\hbar}\right), \tag{8.182}$$

[8] The technicalities of this argument are beyond the scope of this book. However, you can find the details in many quantum mechanics textbooks, e.g. *Introduction to Quantum Mechanics* by B.H. Bransden and C.J. Joachain (Longman, 1989) [8].

where $\lambda \ll 1$ and we would like to calculate the energy of the ground state of this potential. The first thing to note is that the first term is simply an harmonic oscillator potential, for which we already know the solutions (as discussed in Section 4.2). To first order in λ, the ground state energy is given by

$$E_0 \approx E_0^{(0)} + \lambda\, E_0^{(1)}, \tag{8.183}$$

where $E_0^{(0)}$ is the energy of the ground state of the unperturbed potential, i.e. $E_0^{(0)} = \frac{1}{2}\hbar\omega$. The second term is calculated using the result in Eq. (8.176)

$$E_0^{(1)} = \langle \psi_0^{(0)} |\exp\left(-\frac{M\omega\hat{x}^2}{\hbar}\right) |\psi_0^{(0)}\rangle, \tag{8.184}$$

where

$$|\psi_0^{(0)}\rangle = \left(\frac{M\omega}{\pi\hbar}\right)^{1/4} \int_{-\infty}^{\infty} dx \exp\left(\frac{-M\omega x^2}{2\hbar}\right) |x\rangle, \tag{8.185}$$

is the ground state of the unperturbed harmonic oscillator. This means we have

$$\begin{aligned} E_0^{(1)} &= \sqrt{\frac{M\omega}{\pi\hbar}} \int_{-\infty}^{\infty} dx \int_{-\infty}^{\infty} dx' \,\langle x'|\exp\left(\frac{-M\omega x'^2}{2\hbar}\right) \\ &\quad \times \exp\left(-\frac{M\omega\hat{x}^2}{\hbar}\right) \exp\left(\frac{-M\omega x^2}{2\hbar}\right) |x\rangle & (8.186) \\ &= \sqrt{\frac{M\omega}{\pi\hbar}} \int_{-\infty}^{\infty} dx \exp\left(-\frac{2M\omega x^2}{\hbar}\right). & (8.187) \end{aligned}$$

The integral can be performed by using the standard result for the integral of a Gaussian

$$\int_{-\infty}^{\infty} e^{-\gamma x^2}\, dx = \sqrt{\frac{\pi}{\gamma}}. \tag{8.188}$$

This means that (8.187) can be written as

$$E_0^{(1)} = \sqrt{\frac{M\omega}{\pi\hbar}}\sqrt{\frac{\pi\hbar}{2M\omega}} = \frac{1}{\sqrt{2}}. \tag{8.189}$$

The ground state energy of the perturbed potential is, therefore,

$$E_0 \approx \frac{1}{2}\hbar\omega + \frac{\lambda}{\sqrt{2}}, \tag{8.190}$$

to first-order in the perturbation parameter λ. As expected, we recover the unperturbed result in the limit $\lambda \to 0$.

8.16. Exercises

1. Rewrite the state

$$|\psi\rangle = a|0\rangle + b|1\rangle$$

in the basis

$$|\phi_1\rangle = \alpha|0\rangle + \beta|1\rangle$$

$$|\phi_2\rangle = \beta^*|0\rangle - \alpha^*|1\rangle$$

You can assume that $|a|^2 + |b|^2 = 1$ and $|\alpha|^2 + |\beta|^2 = 1$.

2. Show that

$$I = |\uparrow\rangle\langle\uparrow| + |\downarrow\rangle\langle\downarrow|$$

and

$$I = |+\rangle\langle+| + |-\rangle\langle-|$$

are equivalent expressions for the identity operator for a spin-half particle, where the transformation between the basis states is given by equations (8.7) and (8.8).

3. Given a set of states $\{|\psi_i\rangle\}_{i=0}^{n-1}$ that are solutions to the time-independent Schrödinger equation with corresponding energies $\{E_i\}_{i=0}^{n-1}$, i.e.

$$\hat{H}|\psi_i\rangle = E_i|\psi_i\rangle,$$

show that the state

$$|\psi(t)\rangle = \sum_{i=0}^{n-1} \alpha_i\, e^{-iE_it/\hbar}\, |\psi_i\rangle$$

is a solution to the time-dependent Schrödinger equation

$$i\hbar\frac{\partial}{\partial t}|\psi\rangle = \hat{H}|\psi\rangle.$$

4. Suppose the state vectors $|H\rangle$ and $|V\rangle$ respectively denote a horizontally and a vertically polarised photon. Moreover, these states obey the relations $\langle H|H\rangle = \langle V|V\rangle = 1$ and $\langle H|V\rangle = \langle V|H\rangle = 0$. Alternatively, photon states can be described by the vectors, $|R\rangle = (|H\rangle + i|V\rangle)/\sqrt{2}$ and $|L\rangle = (|H\rangle - i|V\rangle)/\sqrt{2}$:

(a) Show that the vectors $|R\rangle$ and $|L\rangle$ are orthogonal.
(b) Show that the vectors $|R\rangle$ and $|L\rangle$ are both normalised.
(c) The general state of a photon is $|\psi\rangle = \alpha|H\rangle + \beta|V\rangle$. An alternative representation is $|\psi\rangle = \gamma|R\rangle + \delta|L\rangle$. Determine γ and δ as a function of α and β.

5. Suppose a particle is prepared in the state $|\psi\rangle = \alpha|1\rangle + \beta|2\rangle + \gamma|3\rangle$ and that the energy operator for this particle is given by $H = E_1|1\rangle\langle 1| + E_2|2\rangle\langle 2| + E_3|3\rangle\langle 3|$, where $|1\rangle$, $|2\rangle$ and $|3\rangle$ form an orthonormal basis.
(a) What are the possible outcomes of an energy measurement?
(b) What is the expectation value for an energy measurement on an ensemble of systems prepared in $|\psi\rangle$?
(c) Calculate the expectation value for an energy measurement on $|\psi\rangle$ if the Hamiltonian is

$$H = \Omega|1\rangle\langle 2| + \Omega|2\rangle\langle 1| + E_1|1\rangle\langle 1| + E_2|2\rangle\langle 2| + E_3|3\rangle\langle 3|.$$

6. The state $|\psi_i\rangle$ is an eigenvector of A with eigenvalue α_i if $A|\psi_i\rangle = \alpha_i|\psi_i\rangle$. Suppose,

$$A = |1\rangle\langle 2| + |2\rangle\langle 1|,$$

where $|1\rangle$ and $|2\rangle$ form an orthonormal basis.
(a) Calculate the two eigenvalues α_i of A.
(b) Calculate the normalised eigenvectors, $|\psi_i\rangle$ of A.

7. Use the condition $a\psi_0(x) = 0$, where a is the annihilation operator, to find the ground state wave function $\psi_0(x)$ of the quantum harmonic oscillator.

8. By direct calculation, show that

$$a^\dagger \psi_1(x) = \sqrt{2}\,\psi_2(x),$$

where $\psi_1(x)$ and $\psi_2(x)$ are respectively the first and second excited states of a quantum harmonic oscillator.

9. Calculate the first order correction to the ground state energy of a particle of mass M trapped in the potential

$$V(x) + \lambda x^3,$$

where $V(x)$ is the harmonic oscillator given by $V(x) = \frac{1}{2}M\omega^2 x^2$ and $\lambda \ll 1$. Comment on the result.

10. Calculate the first order correction to the ground state energy of a particle of mass M trapped in the potential

$$V(x) + \lambda e^x,$$

where $V(x)$ is an infinite square well of length $L = 2\pi$, i.e.

$$V(x) = \begin{cases} 0 & \text{for } 0 \leq x \leq L \\ \infty & \text{otherwise} \end{cases}$$

and $\lambda \ll 1$. You may assume the following integral identity:

$$\int e^x \sin^2(bx)\, dx = \frac{e^x}{2} - \frac{e^x}{1+4b^2}\left(\frac{1}{2}\cos(2bx) + b\sin(2bx)\right).$$

Chapter 9

Second Revolution: Relativity

Unlike quantum physics, relativity does not challenge Boolean logic. According to relativity, objects cannot be in two different states at the same time, but what becomes an issue here is the phrase 'at the same time'. Time itself now becomes a relative concept, dependent on the observer.

The theory of relativity fundamentally changes our ideas of space and time. Newton thought that space and time have an absolute character. He was, however, bothered by the fact that there is no place of absolute rest in his universe where God could sit and guide the world. According to Einstein, neither space nor time are absolute. Only some features of the combined system – spacetime – have an absolute meaning. It is these features that relativity aims to find. The theory of relativity should, therefore, really be called 'the theory of absolutes'.

Suppose that someone asks you when William the Conquerer arrived on the British Isles. If you are like us, and your history knowledge is very poor, then you would require a hint. Say that the person says that it was not later than 1100. But you still may be puzzled. The second hint could be that the date was not before 1000. These two hints may refresh your memory and you would probably guess that it was 1066. This year was before 1100, but it is also after 1000.

Now, remarkable as it may seem, this hint would not help you at all relativistically. Namely, if an event A was not later than the event B, then it does not mean that it was earlier! This is truly amazing. How can this be?

Here is another way of phrasing this amazing fact. Let us use the relationship $A < B$ as A was earlier than B. Now imagine that $C < A$. This, you may think must mean that $C < B$, but this is not true in relativity. The point is simply that time is not linear according to Einstein. So what has actually gone wrong with our everyday (Newtonian) intuition?

The main idea is that Einstein wanted to preserve the law of mechanics that laws of physics are the same for all observers. This was already known to

Galileo and Newton and, in fact, constitutes one of Newton's laws of motion. But Einstein was also aware of the fact that the speed of light is the same for all observers (both experimentally confirmed as well as clear from Maxwell's equations). Let us combine these two to see how time should behave as a consequence of these two rules.

How do we see that some event has taken place? A photon needs to leave the event and arrive at our eye. Then our nervous system gets triggered and our brain processes the information that something has happened. But photons always travel at the speed of light no matter how fast you travel[1]. If you travel at a huge speed and you fire out a photon, then the photon still travels at the speed of light. This can lead to some remarkable conclusions as the following situation (known as Einstein's train) illustrates.

9.1. Simultaneity

Imagine a person standing on a platform at a railway station and observing a fast moving train whizzing by. A person stands in the middle of one of the carriages on the train and fires guns towards the walls at both ends of the carriage. Since he is exactly in the middle and he fires the guns at the same time, he will see the bullets hitting the walls at the same time (since the laws of physics are the same on the train as they are off the train). But, the person standing on the platform will not see these two events happen at the same time. The back wall will move forward as the train is moving and will, therefore, hit the bullet before the other bullet hits the front wall. So, from the perspective of the observer on the platform, the bullets do not hit the walls at the same time. So the notion of simultaneity is not universal. What for you happens at the same time, need not be like that for someone else if they are not stationary with respect to you.

This is an extraordinary conclusion. It means that what we believe to be things that co-exist in time, may for some other observer exist at different times. This led Einstein to famously claim that everything exists at once and any 'change with time is an illusion, albeit a persistent one'. To him this offered great comfort on occasions when he would lose a close friend. This is because the fact that they do not coexist in one frame need not mean that they are not together in some other frame in the relativistic universe. Let us now analyse some other bizarre consequences of Einstein's theory.

9.2. Lorentz Transformations

Einstein postulated two principles from which he then proceeded to derive the rest of relativity. They are:

1. The laws of physics are the same in all inertial frames, or, stating it negatively, this says that you cannot use any experiment to detect that you

[1] This is an assumption of relativity.

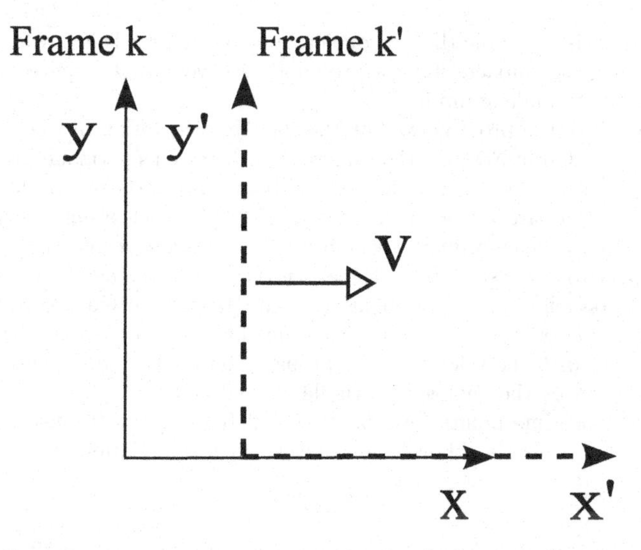

Fig. 9.1. Two Cartesian coordinate frames k and k'. The x-axes of both frames permanently coincide and frame k' moves with velocity v in the x-direction relative to frame k.

are moving uniformly.

2. The speed of light is the same for all inertial observers, i.e. those that moving at uniform speed.

This leads to the following transformations between different inertial observers[2]

$$x' = \gamma(x - vt) \tag{9.1}$$
$$t' = \gamma(t - vx/c^2). \tag{9.2}$$

Here $\gamma = (1 - v^2/c^2)^{-1/2}$ and the speed of light[3] is denoted by c. The spacetime coordinates of an event are (x, t) in one reference frame and (x', t') in another one moving at the speed v relative to the first. The important conclusion is that space and time have no absolute meaning but are observer (or frame) dependent.

The derivation of Lorentz transformations is easy, but a bit lengthy so, if you feel like skipping it, please go ahead. We follow the proof given by Einstein

[2]These transforms had already been worked out in the 1890s by Hendrik Lorentz based on earlier work by George FitzGerald – well before Einstein took up the baton. Einstein's key contribution was to give these results a new and physically relevant interpretation. Lorentz went on to win the Nobel prize in 1902 for providing a theoretical interpretation of the Zeeman effect.

[3]In 1676, the Danish astronomer Ole Rømer was puzzled by the fact that the moons of Jupiter seemed to orbit irregularly. He solved the problem by assuming that the speed of light was finite and, in doing so, was the first person to work out the speed of light.

– the algebra is so simple that anyone can understand it. This is why relativity made such a big impact. It is a profound discovery that requires very little mathematics to understand it.

Here is how the proof goes. For the relative orientation of the co-ordinate systems indicated in Fig. 9.1, the x-axes of both systems permanently coincide. In the present case we can divide the problem into parts by considering first only events that are localised on the x-axis. Any such event is represented with respect to the co-ordinate system k by the abscissa x and the time t, and with respect to the system k' by the abscissa x' and the time t'. A light-signal, which is proceeding along the positive x-axis, is transmitted according to the equation $x = ct$ or $x - ct = 0$. Since the same light signal has to be transmitted relative to k' with the velocity c, the propagation relative to the system k' will be represented by the analogous formula $x' - ct' = 0$.

Those spacetime points (events) that satisfy $x - ct = 0$ must also satisfy $x' - ct' = 0$. Obviously this will be the case when the relation

$$x' - ct' = \lambda(x - ct) \tag{9.3}$$

is fulfilled in general, where λ indicates a constant; for, according to this equation, the disappearance of $(x - ct)$ involves the disappearance of $(x' - ct')$.

If we apply similar considerations to light rays that are being transmitted along the negative x-axis, we obtain the condition

$$x' + ct' = \mu(x + ct), \tag{9.4}$$

where μ is some other constant. By adding (or subtracting) the two equations and introducing for convenience the constants a and b in place of λ and μ where

$$a = \frac{\lambda + \mu}{2} \tag{9.5}$$

and

$$b = \frac{\lambda - \mu}{2} \tag{9.6}$$

we obtain the equations

$$x' = ax - bct \tag{9.7}$$

$$ct' = act - bx. \tag{9.8}$$

We, therefore, have the solution to our problem, so long as we know the value of the constants a and b. These can be found as follows. For the origin of k' we always have $x' = 0$, and hence according to the first equation

$$x = \frac{bc}{a}t. \tag{9.9}$$

If we call v the velocity with which the origin of k' is moving relative to k, we then have

$$v = \frac{bc}{a}. \tag{9.10}$$

The same value v can be obtained from Eq. (9.7), if we calculate the velocity of any other point of k' relative to k, or equivalently the velocity (directed towards the negative x-axis) of a point of k with respect to k'. In short, we can designate v as the relative velocity of the two systems.

Furthermore, the principle of relativity teaches us that, as judged from k, the length of a unit measuring-rod that is at rest with reference to k' must be exactly the same as the length, as judged from k', of a unit measuring-rod that is at rest relative to k. In order to see how the points of the x'-axis appear as viewed from k, we only need to take a snapshot of k' from k. This means that we have to insert a particular value of t (time of k), e.g. $t = 0$. For this value of t we then obtain $x' = ax$ from Eq. (9.7).

Two points of the x'-axis, which are separated by the distance $x' = 1$ when measured in the k' system, are thus separated in our instantaneous photograph by the distance

$$\Delta x = \frac{1}{a}. \tag{9.11}$$

Now, if instead the snapshot is taken from $k'(t' = 0)$, and if we eliminate t from Eq. (9.7), using (9.8), we obtain

$$x' = a\left(1 - \frac{v^2}{c^2}\right)x. \tag{9.12}$$

From this we conclude that two points on the x-axis that are separated by the distance 1 (relative to k) will be represented on our snapshot (from frame k') as being separated by the distance

$$\Delta x' = a\left(1 - \frac{v^2}{c^2}\right). \tag{9.13}$$

But from what has been said, the two snapshots must be identical; hence Δx in (9.11) must be equal to $\Delta x'$ in (9.13), so that we obtain

$$a^2 = \left(1 - \frac{v^2}{c^2}\right)^{-1}. \tag{9.14}$$

The equations (9.10) and (9.14) together determine the constants a and b, i.e. $a = \gamma$ and $b = \gamma v/c$. By inserting the values of these constants into (9.7) and (9.8), we obtain the Lorentz transformations given by (9.1) and (9.2).

How do people in each respective frame measure the x and t coordinates? Well, they need a set of measuring rods (meters) and clocks positioned appropriately. Strictly speaking we need every point in space time to carry a clock and a ruler at the same time. This implies that position and time are dependent on the observer, but the quantity

$$x^2 - c^2t^2 \tag{9.15}$$

has an absolute character. This quantity is observer independent. The proof of this is left as an exercise.

Let us now analyse the rule for the sum of velocities. For that we only need to know that the velocity is defined as the temporal derivative of position, namely $u = dx/dt$. Of course, for the primed observer we have $u' = dx'/dt'$. The question is how are these two velocities related. To answer this we need to look at the infinitesimal version of Lorentz transformations that come directly from Eq. (9.1) and (9.2):

$$dx' = \gamma(dx - vdt) \tag{9.16}$$

$$dt' = \gamma(dt - vdx/c^2) \tag{9.17}$$

and so

$$\frac{dx'}{dt'} = \frac{dx/dt - v}{1 - (v/c^2)dx/dt} \tag{9.18}$$

From this it follows that

$$u' = \frac{u - v}{1 - vu/c^2}, \tag{9.19}$$

which is the relativistic law for addition of velocities.

This is an interesting formula. What it says is that velocities do not just sum up when observers move with respect to one another. Take the following example. Suppose you are standing on a moving escalator and throw a ball in the direction of your motion towards a person who is not on the escalator but standing still on the ground. Say that the speed of the escalator is v and that you throw the ball at the speed u (in the frame of the escalator or your frame, if you like). The simple question is: what is the speed of the ball with respect to the person on the ground? Amazingly enough the answer is not $u + v$. Only if the speed of light were infinite (in which case relativity can be completely ignored) would we recover this result. Otherwise, the result is

$$\frac{u + v}{1 + vu/c^2} \tag{9.20}$$

which can be obtained immediately from (9.19). It is worth convincing yourself that this expression reduces to $u + v$ if the speed of light is very large. This is why, in our everyday world where velocities are much smaller than the speed of light, we intuitively think that the velocities should add as $u + v$.

9.3. Length Contraction

Two amazing implications of the theory of relativity are that length and time are not absolute quantities, but that different observers will record different measurements of the two in their respective reference frames. In the slow moving macroscopic world of humans it is difficult to imagine how length could be a relative concept (much less time which is addressed in the next section). But indeed it is and this is very simple to show from the Lorentz transformations.

Let us consider an object of length L. What is its length as far as a moving observer is concerned? This follows immediately from the Lorentz transformations: if a moving object has length L that means that $\Delta x' = L$. But $\Delta t = 0$

if we are to measure the length of something (in any frame) since our measurements of the two ends must be performed at the same time. This immediately means that in the other frame, where $\Delta x = L'$, we have

$$L' = \frac{L}{\gamma}. \tag{9.21}$$

Since $\gamma > 1$, the moving observer sees the length as being shorter than the length in the object's rest frame – a phenomenon known as length contraction. In other words, the length of a cigar that someone is smoking while moving with respect to you is shorter for you than it is for him (which is probably just as well).

9.4. Time Dilation

In the same vein, time for a moving observer is slower than for someone who is observing it. A pendulum with a period T will appear to have a period

$$T' = \gamma T \tag{9.22}$$

to a moving observer. This also follows from the Lorentz transformations. Therefore, a moving cigar is not only shorter for you, but lasts longer for the moving observer than for you.

Let us derive this result in a different way. Imagine a really simple clock that consists of a beam of light bouncing back and forth between two mirrors (see Fig. 9.2). If the proper distance between the mirrors is L, the time it takes light to travel between the two is just $\Delta t = L/c$. For someone moving at the speed v, however, the light takes

$$\begin{aligned} \Delta t' &= \frac{\sqrt{L^2 + v^2(\Delta t)^2}}{c} = \Delta t\sqrt{1 + \frac{v^2}{c^2}} \\ &\approx \frac{\Delta t}{\sqrt{1 - v^2/c^2}} = \gamma \Delta t, \end{aligned} \tag{9.23}$$

which is the same result as before, i.e. the moving observer sees the clock running slower.

Now comes a really interesting conclusion. Einstein states that no uniform motion can be detected by any means. So, if we are in a moving spaceship, then we cannot tell that we are moving unless we look outside of the window and see that we are moving with respect to planets, for example. This means that all the processes on that ship have to slow down in time with respect to an outside stationary observer in exactly the same way. For example, if you smoke a cigar on the ship, it would last longer. But also, if you were suffering from toothache, this too would last longer (and it would take any dentist longer to fix your tooth as well!). Any difference in the rate of any process would indicate to us that we were moving, and Einstein would have been proven wrong.

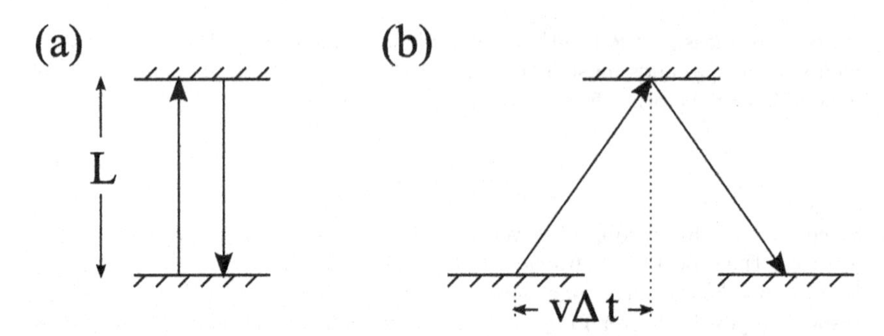

Fig. 9.2. A simple clock consists of light bouncing between two mirrors separated by a proper distance L and takes the proper time $\Delta t = L/c$ to travel from one to the other. The path of the light in the frame of the mirrors is shown in (a). Suppose now that the system is viewed by an observer moving with speed v towards the left. He will observe the light path shown in (b). To the moving observer, the light has to travel the further distance of $\sqrt{L^2 + v^2(\Delta t)^2}$ and so it will take longer to bounce between the mirrors, i.e. the clock will appear to be running slower from his perspective.

Now, this all sounds pretty much like science fiction, but it is not. Time dilation has been tested with very precise atomic clocks, which have been taken on flights around the world and then compared with the clocks that stayed on Earth. And the difference in the final times was exactly as predicted by relativity. But we have an even more astonishing example that relativity is correct. It concerns showers of muons through the atmosphere.

A muon is a type of heavy electron. It has the same charge and spin as an electron and undergoes the same types of interactions. Electrons and muons differ only in their masses – a muon is roughly 200 times heavier than an electron. Because it is so heavy, it is unstable and decays into an electron and two neutrinos (more precisely, a muon-type neutrino and an electron-type antineutrino).

Muons have a short half-life of about $1.56\mu s$. This means that after $1.56\mu s$ there is a 50 percent chance that a given muon has decayed. Ignoring relativistic effects, this means that a bunch of muons travelling at 0.99 times the speed of light ($0.99c$) will travel only about $0.99c \times 1.56\mu s \approx 460$m before half of them have decayed. In the atmosphere, muons are created in a shower at a typical height of 10km and will need 21.7 half-lives of time to reach the ground directly (more if they are coming at an angle). To put this in context, it means that only about one in every 3.5 million muons should reach the Earth's surface.

In reality, a much larger fraction of muons make it through the atmosphere. Relativity tells us that from our point of view, time passes more slowly in a system that moves fast with respect to ourselves. This is also true for our fast-moving muons. While in a laboratory, at rest, muons have a half-life of $1.56\mu s$, they appear to live much longer when they travel at high speed through

the atmosphere. It seems that it is because of relativity that so many muons actually make it down to the earth's surface! This theory has been tested in the laboratory. An experiment carried out at CERN[4] in 1977 accelerated muons to speeds of $0.9994c$ and measured their half-lives in the frame of the laboratory to be about 45.7μs. This is in excellent agreement with the time dilation factor calculated by relativity.
Problem: Given that the speed of muons is $v = 0.99c$, calculate the γ factor. What is the moving muon decay time in our frame? If they get created 10km up in the atmosphere, how long does it take them to reach the earth? What fraction make it to the Earth's surface?
Solution: For muons travelling at $0.99c$, the time dilation factor is about 7.09. Their half-life observed in our ground frame of reference is, therefore, longer by a factor of 7.09 and hence according to relativity the time needed for muons to reach the ground is not 21.7 half-lives but only $21.7/7.09 \approx 3.06$ half-lives. This means that about 1 in 8.3 muons make it to the Earth's surface – a much larger fraction than if we do not account for relativistic effects.

So, to summarise, there are two ways of looking at the (at first sight surprisingly high) rate of muons we observe on Earth. One is to look at it from Earth's reference frame, in which case the half-life of muons is longer than in the rest frame since they are moving at very high speeds. The other reference frame is that of the moving muon, where the distance travelled by the muon is much shorter than that perceived from Earth, by the same γ factor as the dilation of time. This is why the rate of muons is the same independently of which reference frame is used for the calculation.

9.5. The Twin Paradox

The twin paradox is a simple consequence of the time dilation effect and was used by Einstein to illustrate how unusual relativity really is. This effect is so weird that it is called a paradox, although there is really no paradox there – it is a real thing. Suppose that twins are separated and one of them flies off into the space, while the other one stays waiting on Earth for him to return. What happens when the travelling twin returns. According to Einstein, the clock for the twin that was travelling with respect to the one on Earth was ticking at a slower rate. When he got back to Earth, he was, therefore, younger than the stationary twin. How can that be? By symmetry, each twin sees the other one as travelling so each should see the other as aging more slowly. However one twin really does seem to age less than the other. Experiments have been performed where an atomic clock was taken on an aeroplane around the globe and upon returning it was compared to the clock that remained on Earth. The clock on the plane recorded less time and the difference was found to match the prediction of relativity perfectly. There have been a number of explanations of this paradox, all based on the fact that there is no contradiction because,

[4]Bailey J. *et al.* (1977). Measurements of relativistic time dilatation for positive and negative muons in a circular orbit. *Nature* **268**: 301–305 [9].

on closer inspection, the system is not symmetric. One of the twins undergoes acceleration and deceleration, which breaks the symmetry and allows them to age differently.

Let us be a bit more specific about where the difference in the ages of the twins comes from. We start with the fact that the four dimensional spacetime interval is invariant for all observers, which is a direct consequence of the postulates of relativity as we saw above. All we now need to do is calculate the elapsed times for the two twins and a very simple bit of mathematics will show us that one has got to be less than the other.

More generally, suppose someone carrying a high-quality time-piece travels some worldline from event E to event F. 'High-quality' here means that acceleration doesn't affect the time-keeping mechanism. A pendulum clock would not be a good choice[5]. A balance-wheel watch might do ok, a tuning-fork mechanism would be still better, and an atomic clock ought to be nearly perfect. How much time elapses according to the time-piece? In other words what is the proper time, τ, along that worldline between events E and F? Well, simply integrate $d\tau = dt/\gamma$:

$$\tau = \int_E^F \frac{dt}{\gamma} = \int_E^F \sqrt{1 - \frac{v(t)^2}{c^2}}\, dt, \tag{9.24}$$

where

$$v(t) = \left(\frac{dx}{dt}, \frac{dy}{dt}, \frac{dz}{dt} \right) \tag{9.25}$$

is the velocity vector, and $v(t)^2$ is the square of its length:

$$v(t)^2 = \left(\frac{dx}{dt} \right)^2 + \left(\frac{dy}{dt} \right)^2 + \left(\frac{dz}{dt} \right)^2. \tag{9.26}$$

Our integral for the proper time can be difficult to evaluate in general, but certain special cases are a breeze. Let us take two observers Alice and Bob. Bob's event co-ordinates are always $(t, 0, 0, 0)$, so dx, dy, and dz are always 0 for him. So $d\tau$ is just dt, and the difficult integral simply becomes

$$\int_E^F dt, \tag{9.27}$$

i.e. just the difference in the t coordinates. In other words, Bob's elapsed proper time is just the elapsed proper time as measured by any observer in the reference frame in which Bob is at rest. It doesn't stretch credulity too far to suppose that Bob is one of those observers.

[5]The fact that pendulum clocks are affected by acceleration is what meant that they were not suitable for navigation. Accurate time-keeping at sea was crucial for determining the longitude of a ship. Latitude could be determined relatively easily with the aid of a sextant to measure the Sun's angle at noon. In 1714, the British government offered a prize of £20000 (a huge sum) to anyone who could solve this problem. It was eventually solved by John Harrison – a joiner with little formal education – though he had to fight very hard to get the authorities to eventually agree to pay up!

Now how about Alice? For her, dx, dy and dz are not always all zero. So dx/dt, dy/dt and dz/dt are also not always all zero, and their squares, which appear in the formula for $v(t)^2$, are always non-negative, and sometimes positive. So the quantity under the square root (i.e. $1 - v(t)^2/c^2$) is less than or equal to one and sometimes strictly less than one. The only possible conclusion is that the value of Alice's integral is less than that of Bob's integral, i.e. her elapsed proper time is less than Bob's and so she ages less. The exact difference in their aging is given by the time dilation formula. But the argument presented here is instructive as it shows how the invariance of the spacetime interval naturally leads to a difference in their rates of aging.

Therefore there is no paradox here. It is just an uncomfortable feeling generated in our minds because we have evolved in an enviroment where objects move at velocities much less than that of light. We thus take it as intuitively given that all our times have to agree no matter who measures them. But this is wrong and the theory of relativity shows where the mistake is. All this is a nice prelude to an important discussion about causality.

9.6. Causality

Relativity theory is sometimes said to be constructed so that it preserves the causal structure of events in spacetime. What does this mean? We think of one event A as causing another event B if it happens before it and if some signal from A travels to B and initiates its happening. The theory of causality forms a big chunk of philosophy, but here we only want to discuss what relativity has to say about it. We would, therefore, like to talk about ordering of events in spacetime.

Two different events A and B can be related to one another in three different ways:

1. A and B are time-like separated. This means that we can put them in a unique temporal order independently of the frame. One of the two must have happened earlier than the other one. So either $t_A > t_B$ or $t_A < t_B$ for all observers. So, the event where you get up and another one where you make coffee subsequently are time-like separated. No one can perceive them in the opposite order no matter how quickly they are moving. Here we are justified in saying that you getting up somehow causes you making a coffee. And everyone has to perceive it this way, no matter how they travel with respect to you. This is why relativity is said to preserve causality.

2. A and B space-like separated. Here there is no causal connection between the events. For example, it takes light eight minutes to get from the Sun to the Earth. For all that you know, the Sun may have vanished four minutes ago according to your clock, but you won't notice that by looking at the sky, since the Sun will still be there for another four minutes. Your being able to see the Sun and the Sun vanishing are spacelike separated. Different observers will now disagree on the order of these two events, but

we do not care, because neither can cause the other. There is simply no inconsistency in different observers swapping their order in time.

3. A and B are null interval separated. They are connected by a light ray. So, the event where a photon is emitted from the Sun and the event where the photon enters my eye are null separated. Here the two are causally connected.

Fascinatingly, once all events are ordered in the above way, the whole of space-time is fully specified. All we need to do is list all the events that have ever happened or will ever happen, order them with respect to one another according to the above classification and we have the whole relativistic universe. This would take a long time to prove formally and is not really relevant at this stage, but it is an intriguingly simple result nonetheless.

9.7. $E = Mc^2$

In the same way that positions and times are dependent on the observer, so are the notions of energy and momentum. What to one observer will appear to be pure (so called 'rest') energy, will for another one contain a degree of kinetic energy due to motion. Only the combination of the two will have an absolute character independent of the observer.

Energy and momentum also transform in a very similar way to position and time

$$E' = \gamma(E - vp) \tag{9.28}$$

$$p' = \gamma(p - vE/c^2). \tag{9.29}$$

The absolute quantity is

$$E^2 - c^2p^2 \tag{9.30}$$

and it is independent of the observer. All the discussion we had regarding space and time can here be repeated regarding momentum and energy. Each in its own right is a relative quantity (relative to whoever tries to measure it). Again, this is a great surprise, given that for instance we think of energy as an intrinsic property of a given system. But relativity says that it is now a relational property between the system and the observer. This has a stunning consequence regarding the connection between the mass of an object and its energy.

Let us now derive a very famous conclusion that Einstein drew in his first paper on relativity. Imagine that a particle is at rest (i.e. $p = 0$) in one reference frame. What is its energy as far as someone moving at velocity v is concerned? For small velocities, the energy must be whatever it was in the rest frame plus the kinetic energy since the particle now appears to have velocity v. Therefore,

$$E' = E + \frac{M_0v^2}{2}, \tag{9.31}$$

but, relativistically, we also know that

$$E' = \gamma E = E\left(1 + \frac{v^2}{2c^2} + ...\right). \tag{9.32}$$

Equating (9.31) and (9.32) we see that in the rest frame the energy of the particle must be

$$E = M_0 c^2. \tag{9.33}$$

This is Einstein's famous equation showing the equivalence between mass and energy. If we consider the energy in the moving frame, we get

$$E' = \gamma M_0 c^2 = Mc^2,$$

where the mass M in the moving frame is given by

$$M = \frac{M_0}{\sqrt{1-(v/c)^2}}. \tag{9.34}$$

In other words, a moving object gains mass from the perspective of an observer in another inertial frame. In fact, its mass approaches infinity as its velocity approaches c. This is why massive objects cannot be accelerated all the way to the speed of light.

The discussion of the mass-energy equivalence above is not actually how Einstein originally worked it out. It is worth going through his reasoning now both because it is of historical interest and because his simple and characteristically beautiful arguments give a different way of thinking about the effect[6]. Einstein considered a body at rest with mass M. If the body is examined in a frame moving with nonrelativistic velocity v, it is no longer at rest and in the moving frame it has momentum $p = Mv$.

Einstein supposed the body emits two pulses of light with frequency f to the left and to the right, each carrying an equal amount of energy $E/2$. In its rest frame, the object remains at rest after the emission since the two beams are equal in strength and carry opposite momentum. But if the same process is considered in a frame moving with velocity v to the left, the pulse moving to the left will be Doppler red-shifted (i.e. its frequency will decrease) while the pulse moving to the right will be blue-shifted. The blue light carries more momentum than the red light because $p = hf/c$, so that the momentum of the light in the moving frame is not balanced: the light is carrying some net momentum to the right. The object hasn't changed its velocity before or after the emission. Yet in this frame it has lost some right-momentum to the light. The only way it could have lost momentum is by losing mass. Let us now quantify this effect.

The velocity v is small, so the frequency of the right-moving light is shifted by an amount given by the nonrelativistic Doppler shift, i.e. $\Delta f = fv/c$. The

[6]It is always satisfying in physics when the same conclusions can be reached by tackling a problem from a number of different directions.

momentum of the light is its energy divided by c, and it is increased by a factor of v/c. So the right moving light is carrying an extra momentum Δp given by

$$\Delta p = \frac{v}{c}\frac{E}{2c}. \tag{9.35}$$

By a similar argument, the momentum of left-moving light is *reduced* by the same amount, i.e. Δp. So the total right-momentum in the light is twice Δp,

$$2\Delta p = v\frac{E}{c^2}. \tag{9.36}$$

This is the right-momentum that the object lost and given to the light. The momentum of the object in the moving frame after the emission is, therefore,

$$p' = Mv - 2\Delta p = \left(M - \frac{E}{c^2}\right)v. \tag{9.37}$$

So the change in the object's mass is equal to the total energy lost divided by c^2. Since any emission of energy can be carried out by a two step process, where first the energy is emitted as light and then the light is converted to some other form of energy, any emission of energy is accompanied by a loss of mass. Similarly, by considering absorption, a gain in energy is accompanied by a gain in mass. The key outcome of Einstein's analysis is that the mass of a body is a measure of its energy content.

This equation really means that mass and energy are one and the same thing. What is striking in all this is that a particle with a small amount of mass still possess a huge energy content. We can illustrate this fact by looking at the atomic nucleus, an exercise that shaped the world in the last century probably more than any other single event. When two nucleons bind together, the resulting mass is always smaller then the two sum of the individual masses of nucleons. The reason is that the difference in mass is lost to the binding energy (since energy and mass are equivalent) that holds them together. If we reverse this process, then when a heavy atomic nucleus is disintegrated, an amount of energy gets released that is equal to the binding mass times the speed of light squared. If this is done simultaneously with a lot of atoms, the resulting energy could add up to quite a lot, and this is, of course, the basic principle behind the atomic bomb[7]. Horrible though the consequences of the bomb may have been, they do demonstrate one important fact: the sheer capacity of abstract thinking to change the world.

---

[7]Ernest Rutherford, a founding father of nuclear physics, once famously said: "The energy produced by the atom is a very poor kind of thing. Anyone who expects a source of power from the transformation of these atoms is talking moonshine." It is, perhaps, a little unfair to highlight this quote after all of Rutherford's outstanding contributions to atomic physics. However, it illustrates a point we made earlier that when you do something truly revolutionary it is often difficult to foresee all the consequences of your discovery.

9.8. Relativistic Newton's Laws of Motion

Once relativistic laws are acknowledged, they impact on other laws. All of physics in fact has to comply with relativity (even quantum physics as we will see in the next chapter). First let us look at how we need to modify Newton's laws of motion. Newton's second law states that the relationship between force and momentum is

$$F = \frac{d(Mv)}{dt}. \tag{9.38}$$

But Einstein says that the expression for mass has to be modified, therefore,

$$F = \frac{d}{dt}\left(\frac{M_0 v}{\sqrt{1-(v/c)^2}}\right). \tag{9.39}$$

For small velocities the two expressions are very similar, which is why you don't need the relativistic correction to calculate the rocket trajectory on its way to the Moon. And the above equation is really all you need to know to deal with relativistic dynamics.

This logic, for instance, suffices to obtain relativistic corrections to Bohr's atomic spectrum that we calculated in Section 2.5. Remember the calculation there was more or less entirely based on Newton's laws and involved solving the angular momentum and force balance equations

$$n\hbar = M_e v_n r_n \tag{9.40}$$

$$\frac{M_e v_n^2}{r_n} = \frac{Ze^2}{4\pi\epsilon_0 r_n^2}. \tag{9.41}$$

Modifying these equations to take relativity into account, we get

$$n\hbar = \gamma M_e v_n r_n \tag{9.42}$$

$$\gamma\frac{M_e v_n^2}{r_n} = \frac{Ze^2}{4\pi\epsilon_0 r_n^2}. \tag{9.43}$$

We can now redo all our previous calculations from Section 2.5 to obtain the modified formula for the radius of the Bohr orbits as well as their energy[8]. The relativistic radius, for instance, is given by

$$r_n = \frac{a_0 n}{Z}\sqrt{n^2-\alpha^2}, \tag{9.44}$$

where a_0 is Bohr's radius and α is the fine-structure constant given by

$$\alpha = \frac{e^2}{4\pi\epsilon_0\hbar c} \approx \frac{1}{137}. \tag{9.45}$$

Does this make sense? Yes it does. For large values of n we can ignore α^2 (which is small anyway, $\alpha^2 \approx 1/137^2$) and we recover Bohr's old expression

[8]This is set as an exercise at the end of the chapter. Take care when eliminating v_n from the equations to account for the fact that γ is a function of v_n.

$r_n = a_0 n^2/Z$. This is consistent with the fact that, in outer orbits, the electrons are slower (the velocity is inversely proportional to distance) and, therefore, relativity should play a smaller role. The reader is left to work out the expression for energy and analyse it in a similar manner.

Besides spacetime and momentum-energy are there any other conjugate physical quantities? Yes there are. For example, the electric and magnetic fields transform into each other much like space and time. If a collection of charges is stationary with respect to you, you will only register their electric field. However, for an observer moving at certain speed, the charges themselves will be moving and he will, therefore, see an electric current. But the current generates a magnetic field. So electric and magnetic fields are not absolute quantities either, but very much depend on observers.

9.9. General Relativity

Many physicists were seriously puzzled by relativity when Einstein announced it in 1905. It helped a great deal that Planck immediately recognised its merits and started to promote it himself. A number of physicists tried to find mistakes with it and some are still working hard at this. We have seen the twin paradox, which was designed to bring relativity down. Not only did it not undermine relativity, it turned out to be one of its biggest triumphs. One of the most interesting objections, however, came from a Dutch physicist, Paul Ehrenfest. However, rather than uncovering an error, it in fact highlighted the limitations of special theory of relativity (hence the word 'special'). Ehrenfest's objection led to the generalisation of special relativity – to give the theory called general relativity. General relativity contains special relativity as a special case, when acceleration is absent.

Ehrenfest presented the following problem. Suppose that you are an observer standing on a rim of a spinning disk and rotating around with the disk. Now, according to special relativity, the radius of the disk would remain the same as it is perpendicular to your motion, but the circumference would shorten as it is in the direction of your motion (exactly as predicted by Einstein's length contraction). But, wait a minute. How can this be, asked Ehrenfest, when we know from the simple Euclidean geometry that the circumference must be given by $C = 2\pi R$? It shouldn't be the case that the left hand side contracts while the right hand side remains the same.

Ehrenfest thought that his objection dealt a deadly blow to Einstein's relativity. What it did, in fact, is exactly the opposite. It stimulated Einstein to generalise his theory and come up with general relativity. The answer to Ehrenfest's objection is that rotation is not the same as uniform motion. This is because rotation always implies acceleration. A spinning disk indeed contracts in circumference while its radius stays the same! There is only one way this can be possible – the disk itself becomes curved, and in curved geometry $C \neq 2\pi R$. General relativity is all about curvature of space and time for observers moving in the most general possible way including acceleration. The bonus is that gravity is explained as nothing other than the way of talking about this curvature, since gravity is a form of acceleration and all acceleration implies curvature. But this is a topic you will learn about in another book.

Now we have firmly established the theoretical and experimental basis of quantum mechanics as well as relativity, but the two seem to apply to completely different domains of reality. However, there are many circumstances where the two have to be applied together. Can they be merged or are they entirely incompatible? That is the topic of the next chapter

9.10. Exercises

1. Imagine a spaceship floating in space, near Pluto, staying a fixed distance from the Earth. At some time the captain of the ship decides to move out of the Solar system so he turns on the ship's engines and accelerates away. After a while he turns off the engines and the ship continues to coast away from the Earth, now at a constant speed. They coast for a while and then decide to look through their telescope to see how the good old Earth is doing. To their complete horror, they see the destruction of the Earth by a nuclear bomb so powerful that it shatters the Earth into little pieces.

   They realise that the destruction of the Earth is not happening as they watch. It will have taken the light from the explosion some time to travel from the Earth to them. But they are interested in figuring out exactly when the explosion took place. In particular they want to know if explosion took place before or after the captain decided to accelerate away from the Earth.

   The captain argues that the explosion took place right before he turned off the engines. "I had a feeling that the Earth needed us so I cut power", he claims.

   His deputy, on the other hand, argues that the explosion took place right before the captain turned on the engines. "I think that the captain knew the Earth was about to blow up and this is why he decided to accelerate away. He is a traitorous coward and he should be depressurised!", the deputy argues.

   Who is right in this argument?

2. The captain from the previous question gets nervous because there is a chance that the crew will depressurise him. He decides on a preemptive strike and shoots his deputy dead with a laser gun. He fires the gun first, and subsequently the deputy dies - this is what everyone on the ship witnesses.

   However, the captain has the following defence. He says that although on the ship it looks as if he fired first and that is what subsequently killed his deputy, there is (according to relativity) a frame where the deputy died first and then the laser gun was fired. "Since", so the captain argues, "the order of events is completely relative according to Einstein, we can never claim that I am responsible for the death of my deputy. There is always a reference frame in which I am completely innocent."

   Suppose that you are appointed to be the judge in this case. How would you proceed? Is the captain innocent according to relativity?

3. Solve Eqs. (9.42) and (9.43) to find the result given by (9.44) as well as an expression for v_n. Comment on the latter result.

4. This aim of this exercise is to compute the relativistic correction to Bohr's transition energies (wavelengths) in the hydrogen atom.

   Compute the velocity of the electron in the lowest (ground) state of Bohr's model. Calculate thus the true mass of the electron taking its motion into account.

   Use this result to calculate the wavelength of light emitted in the $n = 2$ to $n = 1$ transition. What is the percentage difference between the relativistic and non-relativistic estimates?

5. Suppose that a muon detector at the top of a mountain 2000m high records a flux of 1000 muons per hour. A similar detector located at sea-level near the base of the mountain records 700 muons per hour. Calculate the speed of the muons.

6. When uranium-235 is bombarded with neutrons, it can undergo nuclear fission, whereby the uranium nucleus breaks up into smaller, more-stable nuclei. One such reaction is

$$^{235}\text{U} + \text{n} \longrightarrow {}^{141}\text{Ba} + {}^{92}\text{Kr} + 3\text{n},$$

   where the uranium nucleus breaks up into krypton and barium and releases neutrons in the process. Given the following atomic masses: ^{235}U= 235.043925u; ^{141}Ba=140.914406u; ^{92}Kr=91.926156u; n=1.008665u, find the energy released in the above reaction.

Chapter 10

Relativistic Quantum Mechanics

This chapter is all about putting quantum physics and relativity together. It is not necessarily something that all undergraduate physicists will have to learn, but, we believe that it is important to say a few things about it. For one, the resulting theory - quantum field theory - is the most successful scientific theory ever. Secondly, now that we have got this far, it does not take much more to understand the marriage between the two. Thirdly, unifications are always intellectually pleasing. Why have a patchwork of explanations, when one all-encompassing explanation will do the job. So, here we go.

10.1. Why the Need for Relativistic Quantum Mechanics?

We have seen that all events in physics have to comply with two laws of relativity. Relativity, much like quantum mechanics itself, is a meta theory (to use Einstein's language and classification of theories). This means that whatever other theory we have, it has to comply with relativity (as well as quantum physics). Classical mechanics, for instance, has to be made relativistically invariant. But so does quantum mechanics.

For us so far, the main equation of quantum mechanics has been that of Schrödinger. But this equation is manifestly not relatistically invariant. We can see that space and time enter it in a completely different manner: time through a first derivative and space through a second. Relativity, however, suggests that the two must be put on an equal footing. How is this to be done?

Before we discuss this issue, let us first show how, if the Schrödinger equation were the full description of reality, one would be able to communicate faster than the speed of light. Suppose that Alice has a particle localised inside a box and she wants to use it to communicate with Bob. She either opens the box, thereby letting the particle wave function spread across space or she keeps the box closed. Open and closed constitute one bit of information that Alice can communicate to Bob (0 or 1, yes or no). Since the speed of spreading of the wave function is unlimited (according to Schrödinger's equation) Bob can, by making a measurement, either detect or not detect a particle. This way he receives Alice's message and the communication could be instantaneous. Clearly this contradicts relativity, according to which, no signal can travel faster than light.

Our informal (but precise) argument can be made mathematically rigorous. The first relativisitic equation was the Klein–Gordon one which we now proceed to describe. It was, in fact, first discovered by Schrödinger himself[1], but he discarded it for reasons we will encounter later. The idea of Klein and Gordon was simply to make time a second derivative just like space and this is very natural, since the energy-momentum relativistic relation is quadratic.

10.2. The Klein–Gordon Equation

The Klein–Gordon equation follows the simple rule of substituting $E \to -i\hbar\partial/\partial t$ and $p \to -i\hbar\nabla$ into the relativistic equation $E^2 = p^2c^2 + (Mc^2)^2$. The equation we thus obtain is[2]

$$\left(\frac{1}{c^2}\frac{\partial^2}{\partial t^2} - \nabla^2\right)\Psi + \frac{M^2c^2}{\hbar^2}\Psi = 0. \tag{10.1}$$

By construction, the Klein–Gordon equation is relativistically covariant since it comes from the expression $E^2 = p^2c^2 + (Mc^2)^2$, which itself is relativistically covariant under the energy-momentum Lorentz transformations. The Schrödinger equation therefore features as the low velocity (non-relativistic) limit of the Klein–Gordon equation. As we have mentioned before, when we come up with a more accurate description of natural phemonena, this more accurate description has to contain the previous less accurate one as a special limiting case.

In order to become a bit more familiar with the Klein–Gordon equation, let us first discuss plane wave solutions of this equation. They are as usual given by

$$\Psi = Ne^{i(px-Et)} \tag{10.2}$$

but, because $E = \pm\sqrt{p^2c^2 + (Mc^2)^2}$, we can have negative (as well as positive) energy solutions. This, of course, is in sharp contrast with the Schrödinger equation, which only contains positive energy solutions. Though it would be awkward for a free particle to have negative energy (think negative kinetic energy!), we could still simply ignore this solution as unphysical. However, if we have an interacting particle, it could lose its energy to the environment and cascade down to the most negative solution. This of course simply does not happen. Even if we ignore this last point, there is the related issue of negative probabilities, which will be difficult to interpret and we discuss in the next

[1] Thus conforming to the famous rule of the Russian mathematician Arnold, who said that all laws are named after the wrong people.

[2] ∇ is just a shorthand way of writing the operator $\frac{\partial}{\partial x}\hat{i} + \frac{\partial}{\partial y}\hat{j} + \frac{\partial}{\partial z}\hat{k}$.

section. So, though the Klein–Gordon equation complies with relativity, which is what we set out to achieve, we encounter negative energy solutions that are difficult to interpret.

10.3. Negative Probabilities

There is, as indicated, one problem with the Klein–Gordon equation. Given that energies can be negative and that the probability current is proportional to energy, this means that the probability itself can be negative. This is very difficult to interpret within the standard notion of probability as a relative frequency of positive outcomes. Let us see why this is the case.

The zero (or temporal) component of the probability current is the density and is given by

$$j^0 = \frac{i\hbar}{2Mc^2}\left(\Psi^*\frac{\partial}{\partial t}\Psi - \Psi\frac{\partial}{\partial t}\Psi^*\right). \tag{10.3}$$

For the plane wave, this turns out to be

$$j^0 = \frac{\hbar}{Mc^2}|N|^2 E \tag{10.4}$$

and so we indeed have negative probabilities because E can be negative. There are three other (spatial) components of the current, which do not suffer from the same issue.

Negative probabilities are not necessarily a problem if there is a suitable way of interpreting them. However, if the equation is to refer to a single particle, then negative probabilities are impossible to reconcile with reality. If, on the other hand, we think of the equation as representing many particles (i.e. a field) then we can accommodate negative probabilties. In this case, they will just correspond to positive probabilities but of antiparticles. It is important to stress that the Klein–Gordon equation is obeyed by all quantum particles (irrespective of the aforementioned interpretational issues), no matter what their spin is. This is because it is so simple and general that it does not contain any information about particles' internal structure. If we want something more sophisticated we have to look elsewhere.

10.4. The Dirac Equation

Is there another way of constructing a relativistically covariant Schrödinger equation that does not lead to the Klein–Gordon one? The answer is yes and was first proposed by Dirac. His construction actually resulted in a relativistic equation for an electron. And the bonus was that this equation suggested that there must be a particle identical to the electron, but with the opposite charge – a positron. This turns out to be a general feature: just by requiring that quantum mechanics squares with relativity, we discover that we need antiparticles. Such a surprising conclusion is both profound and beautiful. It is also mysterious. How on earth can just a theoretical fiddling with mathematics lead us to predict the existence of more particles in nature?

Dirac thought that instead of combining energy and momentum in a quadratic fashion, relativity could also be incorporated into quantum mechanics by making the relationship linear. This was at first sight an unusual idea as it is clear that $E = a_x p_x + a_y p_y + a_z p_z + bM$, when squared, cannot reproduce the equation $E^2 = p^2c^2 + M^2c^4$. Dirac's great insight was to see that, if $\alpha_x, \alpha_y, \alpha_z$ were matrices (instead of numbers), then the linear equation can indeed produce the correct relativistic formula.

This is how he came up with his matrices, which, upon closer inspection, need to be four-dimensional. The equation he wrote down was

$$i\hbar\frac{\partial}{\partial t}\Psi = (-i\alpha\nabla + \beta M)\Psi, \tag{10.5}$$

which is the celebrated Dirac equation and the α matrices are known as the Dirac matrices.

Let us again look at the simple plane wave solutions

$$\Psi = \omega e^{ipx}, \tag{10.6}$$

where ω is a four component spinor. Surprisingly, the spin of the electron pops out of this equation even though there was no mention of it in the derivation. To see this, assume we have a particle at rest, i.e. $p = 0$. Then we have four solutions corresponding to the two internal states (spin-up and spin-down) of the electron and the positron

$$u_\uparrow = \begin{pmatrix} 1 \\ 0 \\ 0 \\ 0 \end{pmatrix} \quad u_\downarrow = \begin{pmatrix} 0 \\ 1 \\ 0 \\ 0 \end{pmatrix} \quad v_\uparrow = \begin{pmatrix} 0 \\ 0 \\ 1 \\ 0 \end{pmatrix} \quad v_\downarrow = \begin{pmatrix} 0 \\ 0 \\ 0 \\ 1 \end{pmatrix}. \tag{10.7}$$

The u label corresponds to the electron, the v to positron and ω is in general a superposition of the four possible states.

Computing the current in this case always leads to positive probabilties. Therefore, Dirac's method avoids the need to interpret negative probabilites. However, the energies of particles can still be negative. This potential disaster was converted by Dirac into a huge success that ultimately led to his Nobel prize in 1933. Here is how he reasoned. Suppose that electrons cannot have negative energies, because all the negative energies are already filled by other electrons and, because of Pauli's exclusion principle, no other electrons can exist there. Imagine, however, that one of the negative electrons jumps to a positive energy. This leaves a vacancy in the sea of negative energy. This vacancy has all the properties of an electron, but its charge, although equal in magnitude, has the exact opposite parity. So, we arrive at a positron. In this way, Dirac predicted the existence of antiparticles. The same logic leads us to conclude that each particle has its own antiparticle and in some cases, such as that of photons, the two are one and the same.

The positron was discovered shortly after Dirac's prediction and is one of those amazing stories illustrating the power and beauty of physics. We have some magical theoretical ideas, combined with very creative thinking, implying the existence of a new phenomenon that is then confirmed in practice. Einstein is reported to have said that quantum physics was like black magic. He certainly had a point, it's just that the whole of physics, and not just quantum physics, is like black magic!

10.5. Quantum Field Theory

There is a general way of describing relativistic quantum systems and it goes under the name of quantum field theory. A field is something that is continuous in space and time, i.e. it has some value at every point in space and each instant of time. In this sense, a wave function (whatever equation it obeys – be it Klein–Gordon or Dirac or Schrödinger) is itself a field.

First we briefly review the formalism of quantum field theory. Suppose we have the Schrödinger equation for a single quantum system in one dimension (just for simplicity, since we have been discussing this equation for quite some time; strictly speaking we should be doing this with Klein–Gordon and Dirac, but the methods are absolutely the same[3])

$$\hat{H}(x)\Psi(x,t) = i\hbar\frac{\partial}{\partial t}\Psi(x,t). \tag{10.8}$$

Although this will lead to a non-relativistic field, much of what we say will be true relativistically as well. A formal solution to this equation is

$$\Psi(x,t) = \sum_n b_n(t)\psi_n(x). \tag{10.9}$$

As usual, by substituting this back into the Schrödinger equation, we obtain its time independent version

$$\hat{H}(x)\psi_n(x) = E_n\psi_n(x), \tag{10.10}$$

where E_n are the corresponding energies. In order to convert this into a field equation for many particles, we apply the formalism of second quantisation, which means that we effectively have to 'upgrade' the wave-function Ψ into an operator. A wave function is simply treated as a field (it has a value at every point in space and each instance in time). The name second quantisation implies that we are quantising something that is already quantum, i.e. the Schrödinger equation itself. This procedure, which was also pioneered by Dirac, seems mysterious at first sight. However, all we are doing is making a field comply with Heisenberg's uncertainty principle, i.e. that different components of the field cannot be specified simultaneously. Formally, we write

$$\hat{\Psi}(x,t) = \sum_n \hat{b}_n(t)\psi_n(x). \tag{10.11}$$

[3]...and we will derive the full Hamiltonian to understand Bose–Einstein condensation.

The $\hat{b}_n$ operators are the annihilation operators we introduced in Section 8.13. The conjugate of the above equation becomes

$$\hat{\Psi}^\dagger(x,t) = \sum_n \hat{b}_n^\dagger(t)\psi_n^*(x), \tag{10.12}$$

where $b_n^\dagger$ are the creation operators. Here we only consider the time independent creation and annihilation field operators. Nothing much would change conceptually if the field operators were to be time dependent, although mathematically the whole analysis would become much more involved. The Hamiltonian also has to be 'second quantised' and the new Hamiltonian is given by the average of the old (first quantised) Hamiltonian

$$\tilde{H} = \int dx \hat{\Psi}^\dagger(x,t)\hat{H}(x)\hat{\Psi}(x,t). \tag{10.13}$$

By invoking the orthogonality rules, $\langle\psi_n|\psi_m\rangle = \delta_{mn}$, the second quantised Hamiltonian becomes

$$\tilde{H} = \sum_n E_n \hat{b}_n^\dagger \hat{b}_n \tag{10.14}$$

and this is the same as a set of independent harmonic oscillators. A quantum field (i.e. a second quantised system) is therefore fully equivalent to a bunch of quantum harmonic oscillators. Now we see why we spent a lot of time in this book dedicated to understanding quantum harmonic oscillators. But what if we want to include interactions? Interactions are generally much more difficult to incorporate and lead to all sort of approximate techniques, such as perturbation theory. Here, however, we only want to show how to set the Hamiltonian up.

We now briefly review the second quantisation treatment of interacting systems. The Hamiltonian in the first quantised form is now given by

$$H(x,x') = H_0(x) + V(x,x'), \tag{10.15}$$

where $H_0(x)$ is the free Hamiltonian without interactions and the interactions are described by the term $V(x,x')$. We already know how to second quantise $H_0(x)$, and so we only need to be able to second quantise the interaction Hamiltonian. We will do this now with the intention of obtaining a typical Hamiltonian for the Bose–Einstein gas. This is done by performing the following transformation

$$\tilde{V} = \int dV \hat{\Psi}^\dagger(x,t)\hat{\Psi}^\dagger(x',t)V(x,x')\hat{\Psi}(x,t)\hat{\Psi}(x',t)\,. \tag{10.16}$$

We now need to switch from the creation and annihilation operators in the position picture to the creation and annihilation operators in the momentum picture. After some manipulations and approximations we can arrive at

$$\tilde{H} = \sum_p \frac{p^2}{2M} a_p^\dagger a_p + \sum_{p_1,p_2,q} \tilde{V}(q) a_{p_1+q}^\dagger a_{p_2-q}^\dagger a_{p_2} a_{p_1}. \tag{10.17}$$

So, the first term of the Hamiltonian is (as before) a collection of harmonic oscillators and the second term signifies their interaction. The interaction term tells us that two particles of momenta p_1 and p_2 collide to produce new particles with momenta $p_1 + q$ and $p_2 - q$ respectively. The values of the final momenta are such as to obey the momentum conservation law. $\tilde{V}(q)$ is the strength of interaction which is derived from the original potential $V(x, x')$. The exact form of this is unimportant. In fact, it is sometimes customary to assume that the interaction is independent of the momentum change q and to write (after omitting the tilde symbol)

$$H = \sum_p \frac{\hbar p^2}{2m} a_p^\dagger a_p + V \sum_{p_1, p_2, q} a_{p_1+q}^\dagger a_{p_2-q}^\dagger a_{p_2} a_{p_1} \,. \tag{10.18}$$

This becomes the same as the Hamiltonian for a degenerate, but almost ideal Bose gas[4]. The creation and annihilations operators, of course, obey the usual bosonic commutation relations[5]. In order to solve and find the eigenvalues of this Hamiltonian we need to convert it into a collection of non-interacting harmonic oscillators (or modes). The transformations that achieve this are called the Bogoliubov transformations and are given by

$$a_p = u_p b_p + v_p b_{-p}^\dagger \tag{10.19}$$

$$a_p^\dagger = u_p b_p^\dagger + v_p b_{-p}, \tag{10.20}$$

where the coefficients u and v (which are functions of the momentum p) have to be chosen so that the Hamiltonian for b modes has no interacting parts. The resulting Hamiltonian, written in terms of the b modes, is given by

$$H = E_0 + \sum_{p \neq 0} E(p) b_p^\dagger b_p, \tag{10.21}$$

where E_0 is, for us, an unimportant constant and $E(p)$ is the energy of the new modes[6]. Here we have a collection of decoupled (i.e. non-interacting) modes as before.

We can see the effect of the interaction in the following way. The ground state written in terms of the a modes is given by

$$|\psi_0\rangle = \prod_{p \neq 0} \frac{1}{u_p} \sum_{i=0}^{\infty} \left\{ -\frac{v_p}{u_p} \right\}^i |n_{-p} = i; n_p = i\rangle. \tag{10.22}$$

---

[4]We have discussed Bose condensation before, but in a very simple de Broglie wave picture. Quantum field theory, however, is how it is done properly.

[5]If we started with the full Klein–Gordon field, we would also obtain bosons upon second quantisation. This is an instance of the spin statistics theorem, which states that particles with integral spin, in this case zero, are bosonic in their statistical properties.

[6]All the details, which are not important for the analysis here, can be found, for example, in *Statistical Physics Part 2: Landau and Lifshitz Course of Theoretical Physics* by E.M Lifshitz and L.P. Pitaevskii (Reed, 1980) [10].

We see from this that we can no longer treat the p and $-p$ modes as independent – their states are intrinsically correlated. Such a state is called an entangled state and the entanglement generated here is a direct consequence of the interactions. Quantum entanglement has many fundamental properties that exemplify the bizarre nature of quantum mechanics, but it also has a number of applications that are relevant for modern technology. We will explore this subject in more detail in the next chapter.

10.6. Outlook

Quantum field theory has been an extremely successful enterprise. It is the most accurate description of nature we have and it has led to a unification of three of the fundamental forces (electromagnetic, strong and weak). It has revealed many important features of the Universe to us including the fact that particles fundamentally get created and destroyed, every particle has an antiparticle, and that spin is an intrinsic property of every particle and is directly connected with its statistics.The fact that half integral spins are fermions and full integral spins bosons is ultimately only explicable within quantum field theory. However, quantum field theory also has its downsides. One of the most pressing open problems is how to apply it to gravity. All attempts to quantise gravity have so far been unsuccessful. This may present us with three different crudely divided scenarios: quantum field theory fails and has to be modified in the light of gravity (which itself is another field, but classical), or gravity fails and ultimately has to be quantised (this is the direction followed by most researchers), or gravity is just a product of quantum noise and need not be quantised. It will certainly be fascinating to see how this field develops in the fullness of time.

10.7. Exercises

1. Why is a single mode of the field equivalent to a unit-mass harmonic oscillator? What is a photon according to this picture?

2. In terms of creation and annihilation operators the Hamiltonian for a single-mode field of frequency ω is

$$H = \hbar\omega\left(\hat{a}^\dagger\hat{a} + \frac{1}{2}\right).$$

   A coherent state of this field with the amplitude α is given by

$$|\alpha\rangle = e^{-|\alpha|^2/2}\sum_n \frac{\alpha^n}{\sqrt{n!}}|n\rangle.$$

   What is the probability of obtaining n photons in the field? What is the average energy of the field in this state in terms of α?

3. In light of the previous question, what does $|\alpha|^2$ represent physically?

4. Solve the Schrödinger equation to obtain the free evolution of this state. What happens to the amplitude α during the evolution? How does the probability of observing n photons evolve in time?

5. Finally, let us finish this section with a challenging problem that draws together many of the ideas we have learnt about in the book so far including relativity, perturbation theory and the structure of the hydrogen atom. If you can solve this problem, or at least follow the answer provided at the back of the book, then you have truly mastered the ideas we set out to convey.

   In atomic physics, the fine-structure describes the splitting of the spectral lines due to relativistic corrections. In Chapter 5 we calculated the ground state of the hydrogen atom by finding the lowest energy solution of the Schrödinger equation with the Hamiltonian

$$H = \frac{p^2}{2M} + V(r),$$

   where

$$V(r) = -\frac{e^2}{4\pi\epsilon_0 r}.$$

   The first term in the Hamiltonian is the (non-relativistic) kinetic energy. Use perturbation theory to find the first order correction to the ground state energy when the relativistic form of the kinetic energy is used instead. Recall that the relativistic form of the kinetic energy is

$$T = \sqrt{p^2c^2 + M^2c^4} - Mc^2,$$

where the first term is the total relativistic energy and the second term is the rest energy of the electron. You may also find the following integral identity useful:

$$\int xe^{\alpha x}\, dx = e^{\alpha x}\left(\frac{x}{\alpha} - \frac{1}{\alpha^2}\right).$$

Chapter 11

Quantum Entanglement

Perhaps the most fundamental feature that distinguishes quantum physics from its classical counterpart is a property called quantum entanglement. In classical mechanics particles can be correlated over long distances simply because one observer can prepare a system in a particular state and then tell a different observer to prepare the same state. However, all the correlations generated in this way can be understood perfectly well using classical probability distributions and classical intuition. In quantum mechanics things are very different and we can prepare two particles in such a way that the correlations between them cannot be explained classically. Such states are called entangled states.

In this chapter we will introduce the basic ideas and important issues surrounding quantum entanglement and discuss some of its consequences, such as teleportation and its use in quantum computation. This is a large and active field of current research and here we can only give a brief overview and highlight a few selected applications[1]. We hope, however, that this will pique the reader's interest and further demonstrate just how intriguing, baffling and fascinating quantum mechanics continues to be even a hundred years after its discovery.

In many ways this final chapter brings us full circle. We started the book by showing how the early fathers of quantum mechanics applied ideas from thermodynamics to show that energy must come in discrete quanta. We then showed how quantum theory was developed and introduced special relativity. By combining quantum mechanics with relativity we were led to quantum field theory and the idea of entanglement. In this chapter, we explore entanglement in more detail and will see that the fundamental laws that govern entanglement transformations take us right back to ideas at the very heart of thermodynamics[2].

[1]For a much more comprehensive review of the subject, we refer the reader to *Quantum Computation and Quantum Information* by M. Nielsen and I. Chuang (Cambridge, 2000) [11] or *Introduction to Quantum Information Science* by V. Vedral (Oxford, 2006) [12].

[2]This chapter is based on the article Plenio M., Vedral V. (1998). Teleportation, entanglement and thermodynamics in the quantum world. *Contemporary Physics* **39**: 431–446 [13].

11.1. What is Entanglement?

We should perhaps start by being a bit more precise about what we mean by entanglement. Entanglement is a property of two or more quantum systems that exhibit correlations that cannot be explained by classical physics. It is a key resource in lot of quantum applications such as quantum computing, quantum cryptography, and teleportation. We can understand it better by considering a particular example.

States of two quantum systems can be considered together by taking their tensor product. For example, the two states $|+\rangle$ and $|-\rangle$ together give

$$|+\rangle|-\rangle = \frac{1}{2}(|0\rangle + |1\rangle)(|0\rangle - |1\rangle) \tag{11.1}$$

$$= \frac{1}{2}(|00\rangle - |01\rangle + |10\rangle - |11\rangle). \tag{11.2}$$

For notational simplicity, we have written states like $|0\rangle|1\rangle$ as $|01\rangle$ and will continue this convention in the remainder of the chapter. Well-known examples of entangled states are the Bell states defined by

$$|\Phi^+\rangle = (|00\rangle + |11\rangle)/\sqrt{2} \tag{11.3}$$

$$|\Phi^-\rangle = (|00\rangle - |11\rangle)/\sqrt{2} \tag{11.4}$$

$$|\Psi^+\rangle = (|01\rangle + |10\rangle)/\sqrt{2} \tag{11.5}$$

$$|\Psi^-\rangle = (|01\rangle - |10\rangle)/\sqrt{2}. \tag{11.6}$$

The last of these, $|\Psi^-\rangle$, is often called a singlet. Let us try and express the first one, $|\Phi^+\rangle$, as a combination of two qubits[3]: $\alpha|0\rangle + \beta|1\rangle$ and $\gamma|0\rangle + \delta|1\rangle$. By expanding this expression, we can find the values of α, β, δ and γ

$$(|00\rangle + |11\rangle)/\sqrt{2} = |\Phi^+\rangle = (\alpha|0\rangle + \beta|1\rangle)(\gamma|0\rangle + \delta|1\rangle) \tag{11.7}$$

$$= \alpha\gamma|00\rangle + \alpha\delta|01\rangle + \beta\gamma|10\rangle + \beta\delta|11\rangle. \tag{11.8}$$

Equating coefficients we get $\alpha\gamma = \beta\delta = 1/\sqrt{2}$ and $\alpha\delta = \beta\gamma = 0$. The latter expression implies that either β or γ is zero, but this is inconsistent with the former expression, i.e. $\alpha\gamma = \beta\delta = 1/\sqrt{2}$. Therefore we conclude that the Bell state $|\Phi^+\rangle$ cannot be broken down into two separate qubit states – in other words, it is entangled. A similar argument can be applied to the other three Bell states as well.

More generally, we say that a pure quantum state is entangled across two or more subsystems when it cannot be expressed as a tensor product state in those systems. We can also define entanglement for mixed states. Suppose the mixed state, ρ, can be written on the partition $1, 2, \cdots N$ as

$$\rho = \sum_i p_i \, \rho_i^{(1)} \otimes \rho_i^{(2)} \otimes \cdots \rho_i^{(N)}, \tag{11.9}$$

[3] A two level system in quantum mechanics is also called a quantum bit or qubit, in direct analogy with the classical bit of information, which is just two distinguishable states of some system. Unlike the classical bit, a qubit can be in a superposition of its basis states.

where $\rho_i^{(j)}$ is a state on the subensemble j, and all the coefficients p_i are classical probabilities, i.e. they are all positive semidefinite and sum to unity. In that case, the state ρ is not entangled – another way of saying this is that it is a separable state. Otherwise, if the state ρ cannot be written in the form of (11.9) then, by definition, it is entangled on the partition $1, 2, \cdots N$.

Entanglement can exist between two subsystems that are separated by great distances. We will see shortly how this enables us to perform interesting quantum information processing tasks, such as teleporting quantum states. Before we do that, however, let us consider some of the consequences of entanglement and why it is so different from anything that can be described classically.

11.2. Bell's Inequalities

In 1935, Einstein Podolsky and Rosen (EPR) developed a thought experiment that they claimed demonstrated the incompleteness of quantum mechanics. What they meant was that there were certain things that quantum mechanics would not be able to describe. This would, of course, be very bad news for quantum mechanics.

Their basic argument was as follows. Suppose we have two particles in the entangled Bell state[4],

$$|\Phi^+\rangle = \frac{1}{\sqrt{2}}\left(|0_A\rangle|0_B\rangle + |1_A\rangle|1_B\rangle\right) \tag{11.10}$$

and that the two parties (Alice and Bob) corresponding to the two qubits (labelled A and B) are separated by many light years. Now the strange thing is that a measurement made by one party seems to instantaneously affect the qubit held by the other party. If Alice measures her particle to be $|0\rangle$, then so will Bob, and similarly for $|1\rangle$. How can this be since we have already seen from relativity that no information can travel faster than the speed of light? Not surprisingly, Einstein was not comfortable with this effect and called it 'spooky action at a distance'. He claimed that, if we are not to completely abandon relativity, then we must conclude that our quantum description of the system is incomplete. To put it another way, if we cannot accept that what Alice does can instantaneously affect Bob (who is several light years away), then there must be something more to the quantum formalism that we are missing. This missing part, it was claimed, were local 'hidden variables'. These were variables that did not appear in the present formalism (hence hidden) but, it was claimed, predetermined the measurement outcomes. So, it was not the fact that, by measuring her qubit as $|0\rangle$, Alice also caused Bob's to become $|0\rangle$, but rather it was predetermined that Alice measuring $|0\rangle$ meant that Bob would also.

The hidden variable hypothesis remained controversial for many years until 1964 when John Bell[5] proposed an experiment for testing it. He found various

[4]Such a state is often called an EPR pair.

[5]After whom the Bell states are named.

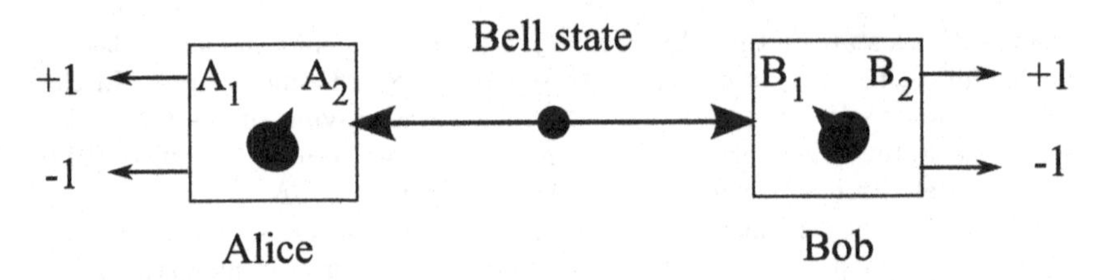

Fig. 11.1. EPR experiment. Alice and Bob each receive one qubit from a Bell state and make measurements on it. They each have an apparatus that has two settings (A_1 and A_2 for Alice and B_1 and B_2 for Bob) and they choose these settings randomly and independently of each other. For either setting, there are two measurement outcomes, $+1$ and -1. By recording their measurement results and seeing how they correlate with the other party's, Alice and Bob are able to determine whether a local hidden variable is governing the outcomes.

inequalities that any local hidden variable model would have to satisfy. Experimental tests were carried out using these inequalities and violations were found. In other words, the local hidden variable hypothesis was invalidated. It seems that the 'spooky action at a distance' (more technically called nonlocality) so disliked by Einstein really is a genuine feature of the quantum world[6].

Let us now spend some time reviewing Bell's inequalities in a bit more detail. We can do this by imagining the experiment shown in Fig. 11.1, which is more or less what Bell did. Alice and Bob each have one particle from a pair that was created in some quantum experiment, the details of which are unimportant. Alice and Bob also each have two sets of measuring apparatus and they agree that they will both simultaneously and independently choose one apparatus and use it to make a measurement of their particle. Let us call Alice's apparatuses A_1 and A_2 and Bob's B_1 and B_2. For convenience, we say that each measurement has two possible outcomes $+1$ and -1. Bell's inequality then states that for any local hidden variable theory, the following must be obeyed,

$$|E(A_1B_1) + E(A_1B_2) + E(A_2B_1) - E(A_2B_2)| \leq 2, \tag{11.11}$$

where $E(A_iB_j)$ is the expectation value of the measurements A_i and B_j.

Bell made two key assumptions in his formulation:

1. Each measurement reveals an objective physical property of the system. This means that, just like in classical physics, the particle had this property before the measurement was made even though it might not be known to us.

[6]We should add that it turns out that this result does not invalidate Einstein's relativity after all. Correctly stated, relativity says that no *information* can travel faster than the speed of light. Although Alice's actions can instantaneously affect Bob's qubit, a careful analysis shows that this cannot be used to convey any information. Since it does not constitute an instant messaging service, the theory of relativity remains intact.

2. A measurement made by Alice cannot affect the outcome of Bob's measurement and vice versa. This follows from the theory of relativity which stipulates that no signal can travel faster than the speed of light.

To prove Bell's inequality, we write the left hand side of (11.11) as

$$\begin{aligned} &|E(A_1B_1) + E(A_1B_2) + E(A_2B_1) - E(A_2B_2)| \\ &\qquad = |E(A_1B_1 + A_1B_2 + A_2B_1 - A_2B_2)| \\ &\qquad = |E(A_1(B_1 + B_2) + A_2(B_1 - B_2))| \,. \end{aligned} \tag{11.12}$$

The outcome of each experiment is ± 1, which leads to two cases:

1. $B_1 = B_2$, which means that $A_1(B_1 + B_2) + A_2(B_1 - B_2) = A_1(2B_1) = \pm 1(\pm 2) = \pm 2$.

2. $B_1 = -B_2$, which means that $A_1(B_1 + B_2) + A_2(B_1 - B_2) = A_2(-2B_2) = \pm 1(\mp 2) = \pm 2$.

So, in either case we have $A_1B_1 + A_1B_2 + A_2B_1 - A_2B_2 = \pm 2$. This means that, averaging over many measurements, the expectation value must be bounded by

$$-2 \leq E(A_1B_1 + A_1B_2 + A_2B_1 - A_2B_2) \leq +2. \tag{11.13}$$

Taking the absolute value gives us Bell's inequality (11.11).

We know that Bell's inequality must be obeyed by all states that are governed by local hidden variables and, as discussed above, many experiments have been performed that show violations of it. For example an EPR pair can give values up to $|E(A_1B_1) + E(A_1B_2) + E(A_2B_1) - E(A_2B_2)| = 2\sqrt{2}$ when maximised over the detector settings.

With the experimental demonstration of the violation of Bell's inequality, it seemed that the question of the non-locality of quantum mechanics had been settled once and for all. However, in recent years it turned out this conclusion was premature. While the entanglement of pure states can be viewed as well understood, the entanglement of mixed states still has many properties that are mysterious, and in fact new problems keep appearing. The reason for the problem with mixed states lies in the fact that the quantum content of the correlations is hidden behind classical correlations in a mixed state. One might expect that it would be impossible to recover the quantum content of the correlations but this conclusion would be wrong. Special methods have been developed that allow us to 'distill' out the quantum content of the correlations in a mixed quantum state. In fact, these methods showed that a mixed state which does not violate Bell inequalities can nevertheless reveal quantum mechanical correlations, as one can distill from it pure entangled states that violate Bell inequalities. Therefore, Bell inequalities are not the last word in the theory of quantum entanglement.

11.3. Quantum Teleportation

One of the most fascinating and bizarre consequences of entanglement is that it allows us to teleport quantum states! Generally when we think of teleportation, we think of an object positioned at a place A and time t that first 'dematerializes' and then reappears at a distant place B at some later time $t+T$. Quantum teleportation implies that we do the same thing but to a quantum object. In reality, quantum teleportation is a bit different because we do not teleport the whole object but just its state from particle A to particle B. As quantum particles are indistinguishable anyway, this amounts to 'real' teleportation.

One way of performing teleportation (and certainly the way portrayed in various science fiction movies such as *The Fly*) is first to learn all the properties of that object, thereby possibly destroying it. This information is then sent as a classical string of data to B where another object with the same properties are re-created. One problem with this picture is that, if we have a single quantum system in an unknown state, we cannot determine its state completely because of the uncertainty principle. More precisely, we would need an infinite ensemble of identically prepared quantum systems to be able to determine its quantum state completely. So it seems that the laws of quantum mechanics prohibit teleportation of single quantum systems. However, the very feature of quantum mechanics that leads to the uncertainty principle (the superposition principle) also allows the existence of entangled states. These entangled states are the key to teleportation.

It turns out that there is no need to learn the state of the system in order to teleport it. However, as we shall see, we do need to send some of the information classically[7], e.g. by phone. After the teleportation is completed, the original state of the particle at A is destroyed (although the particle itself remains intact) and so is the original entanglement in the quantum channel. Quantum teleportation is not just some mathematical quirk. Strange as it may seem, it is a real effect and has already been implemented experimentally using single photons and ions.

Let us now describe quantum teleportation in the form it was originally proposed by Charles Bennett and coworkers in 1993 and shown in Fig. 11.2. Suppose that Alice and Bob, who are distant from each other, wish to implement a teleportation procedure. Initially they need to share a maximally entangled pair of qubits, such as the Bell state,

$$|\Psi_{AB}\rangle = (|0_A\rangle|0_B\rangle + |1_A\rangle|1_B\rangle)/\sqrt{2}\,, \tag{11.14}$$

where the first ket (with subscript A) belongs to Alice and second (with subscript B) to Bob.

Now suppose that Alice receives a qubit in an unknown state (let us label it $|\Phi\rangle = a|0\rangle + b|1\rangle$) and she has to teleport it to Bob. The state has to be unknown to her because otherwise she can just phone Bob up and tell him all the details

[7]It is this step that means teleportation cannot violate special relativity, since the classical information that needs to be sent cannot travel faster than the speed of light.

of the state, and he can then recreate it on a particle that he possesses. If Alice does not know the state, then she cannot measure it to obtain all the necessary information to specify it. Therefore, she has to resort to using the state $|\Psi_{AB}\rangle$ that she shares with Bob. To see what she has to do, we write out the total state of all three qubits

$$|\Phi_{AB}\rangle = |\Phi\rangle|\Psi_{AB}\rangle = (a|0\rangle + b|1\rangle)(|00\rangle + |11\rangle)/\sqrt{2}. \tag{11.15}$$

However, the above state can be written in the following convenient way[8]

$$\begin{aligned} |\Phi_{AB}\rangle &= \frac{1}{\sqrt{2}}(a|000\rangle + a|011\rangle + b|100\rangle + b|111\rangle) \\ &= \frac{1}{2}\left[|\Phi^+\rangle(a|0\rangle + b|1\rangle) + |\Phi^-\rangle(a|0\rangle - b|1\rangle)\right. \\ &\quad \left. + |\Psi^+\rangle(a|1\rangle + b|0\rangle) + |\Psi^-\rangle(a|1\rangle - b|0\rangle)\right], \end{aligned} \tag{11.16}$$

where the Bell states $|\Phi^+\rangle$, $|\Phi^-\rangle$, $|\Psi^+\rangle$ and $|\Psi^-\rangle$ given by (11.3)–(11.6), form an orthonormal basis of Alice's two qubits (remember that the first two qubits belong to Alice and the last qubit belongs to Bob). This is a very useful way of writing the state of Alice's two qubits and Bob's single qubit, because it displays a high degree of correlations between Alice's and Bob's parts: to every state of Alice's two qubits, i.e. $|\Phi^+\rangle, |\Phi^-\rangle, |\Psi^+\rangle, |\Psi^-\rangle$, corresponds a state of Bob's qubit. In addition the state of Bob's qubit in all four cases looks very much like the original qubit that Alice has to teleport to Bob. It is now straightforward to see how to proceed with the teleportation protocol (see Fig. 11.2):

1. Upon receiving the unknown qubit in state $|\Phi\rangle$ Alice performs projective measurements on her two qubits in the Bell basis. This means that she will obtain one of the four Bell states randomly and with equal probability.

2. Suppose Alice obtains the state $|\Psi^+\rangle$. Then the state of all three qubits collapses to the following state

$$|\Psi^+\rangle(a|1\rangle + b|0\rangle), \tag{11.17}$$

where the last qubit belongs to Bob as usual. Alice now has to communicate the result of her measurement to Bob over the phone, for example. The point of this communication is to inform Bob how the state of his qubit now differs from the state of the qubit Alice was holding previously.

3. With this information Bob now knows exactly what he has to do in order to complete the teleportation. He has to apply a unitary transformation on his qubit, which simulates a logical NOT operation: $|0\rangle \to |1\rangle$ and $|1\rangle \to |0\rangle$. He thereby transforms the state of his qubit into the state $a|0\rangle + b|1\rangle$, which is precisely the state that Alice had to teleport to him

[8]Here we are only rewriting the expression in a different basis, and there is no physical process taking place in between.

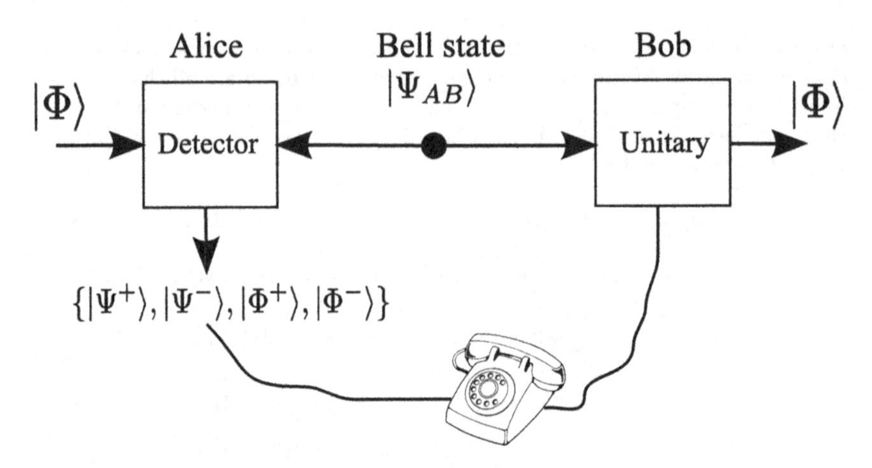

Fig. 11.2. The teleportation protocol. Alice and Bob each possess one qubit from the Bell state, $|\Psi_{AB}\rangle$ and Alice also has the qubit $|\Phi\rangle$ that she wishes to teleport to Bob. To achieve this, Alice makes a joint measurement on her two qubits, i.e. $|\Phi\rangle$ and her part of the Bell state, in the Bell state basis. She obtains one of four possible results, i.e. $|\Psi^+\rangle$, $|\Psi^-\rangle$, $|\Phi^+\rangle$, or $|\Phi^-\rangle$ and communicates the outcome of her measurement to Bob by classical means, e.g. over the phone. Based upon this measurement outcome, Bob makes one of four local unitary operations on his qubit to obtain the state the Alice originally possessed, i.e. $|\Phi\rangle$.

initially. This completes the protocol. It is easy to see that if Alice obtained some other Bell state then Bob would have to apply some other simple operation to complete teleportation. We leave it to the reader to work out the other two operations (note that if Alice obtained $|\Phi^+\rangle$ Bob would not have to do anything).

An important fact to observe in the above protocol is that all the operations (Alice's measurements and Bob's unitary transformations) are *local* in nature. This means that there is never any need to perform a *global* transformation or measurement on all three qubits simultaneously, which is what allows us to call the above protocol a genuine teleportation. It is also important that the operations that Bob performs are independent of the state that Alice tries to teleport.

After teleportation, the initial state is destroyed, i.e. it becomes a maximally mixed state of the form $(|0\rangle\langle 0| + |1\rangle\langle 1|)/2$. This has to happen since otherwise we would end up with two qubits in the same state at the end of teleportation (one with Alice and the other one with Bob). So, effectively, we would clone an unknown quantum state, which is known to be impossible by the laws of quantum mechanics[9]. We also see that at the end of the protocol the quantum

[9] We can understand why quantum states cannot be cloned using a simple argument based on the uncertainty principle. If we could clone a quantum state then we would be able to measure the position on one copy and the momentum on the other. This suggests that we could know both simultaneously – a fact that is forbidden by the uncertainty principle.

entanglement of $|\Psi_{AB}\rangle$ is completely destroyed. Does this have to be the case in general or might we save that state at the end by perhaps performing a different teleportation protocol? Could we, for example, have a situation where Alice teleports a quantum state from to Bob and afterwards the quantum channel is still preserved? This would be of great practical advantage, because we could use a single entangled state over and over again to teleport an unlimited number of quantum states from Alice to Bob. Unfortunately the answer to the above question is no: the entanglement of the quantum channel has to be destroyed. The analytical proof of this is hard. However, the rest of this chapter introduces new ideas and principles that will allow us to explain more easily why this needs to be so. First, however, we need to understand why entanglement is necessary for teleportation in the first place.

11.4. Why is Entanglement Necessary?

Quantum teleportation does not work if Alice and Bob share a separable (disentangled) state. For example, if we take $|\Psi_{AB}\rangle = |00\rangle$ and run the same protocol as above, then Bob's particle always remains unchanged, i.e. there is no teleportation. In this case the total state of the three qubits would be

$$|\Phi_1\rangle = (a|0\rangle + b|1\rangle)|00\rangle \ . \tag{11.18}$$

We see that whatever Alice does to the first two qubits, Bob's qubit will always be in the state $|0\rangle$. It is thus completely uncorrelated to Alice's two qubits and no teleportation is possible.

One might be tempted to say that teleportation is unsuccessful because there are no correlations between A and B, i.e. A and B are statistically independent from each other. So, let us therefore try a state of the form

$$\rho_{AB} = \frac{1}{2}\left(|00\rangle\langle 00| + |11\rangle\langle 11|\right) \ . \tag{11.19}$$

This state is a statistical mixture of the states $|00\rangle$ and $|11\rangle$, both of which are separable. This is equivalent to Alice and Bob sharing either $|00\rangle$ or $|11\rangle$, but being completely uncertain about which state they have. This state is clearly correlated, because if Alice has 0 so does Bob, and if Alice has 1 so does Bob. However, since both the states are separable and neither one of them achieves teleportation then their mixture cannot do it either. The interested reader can convince himself of this fact by actually performing the necessary calculation, which is messy but straightforward. It is important to stress that Alice is in general allowed to perform any measurement on her qubits and Bob any state-independent transformation on his qubit, but the teleportation would still not work with the above state. In fact, it follows that if $\{|\alpha_A^i\rangle\}$ is a set of states belonging to Alice and $\{|\beta_B^j\rangle\}$ a set of states belonging to Bob, then the most general state that cannot achieve teleportation is of the form

$$\sigma_{AB} = \sum_{ij} p_{ij} |\alpha_A^i\rangle\langle\alpha_A^i| \otimes |\beta_B^j\rangle\langle\beta_B^j| \,, \tag{11.20}$$

where p_{ij} are a set of probabilities such that $\sum_{ij} p_{ij} = 1$. This is just a way of writing the general separable state of two qubits (see Eq. 11.9). This state might contain some classical correlations as we have seen above, but any form of quantum correlation, i.e. entanglement, is completely absent. This establishes an important fact: entanglement plays a key role in the manipulation of quantum information.

11.5. The Non-Increase of Entanglement under Local Operations

Our discussion above leads us directly to one of the central laws of quantum information processing. This law encapsulates the fact that if Alice and Bob share no entanglement they cannot achieve teleportation. It can be stated as:

1. The fundamental law of quantum information processing:
Alice and Bob cannot, with no matter how small a probability, turn a separable state σ_{AB} into an entangled state by local operations and classical communication.

The gist of the proof relies on *reductio ad absurdum.* Suppose they could turn the separable state σ_{AB} into an entangled state by local operations and classical communication. They could then use the entangled state they obtained for teleportation. Thus in the end it would be possible to teleport using separable states, which contradicts our discussion in the previous section. Note the clause in the fundamental law that says 'with no matter how small a probability'. This is, of course, very important to stress as we have seen that teleportation is never possible with separable states.

This fundamental law can also be restated in a more general form. We have seen that non-local features, i.e. entanglement, cannot be created by acting locally. This implies that if Alice and Bob share a certain amount of entanglement[10] initially, they cannot increase it by only local actions and classical communication. So we can now restate the fundamental law in the following, more general, way.

2. The fundamental law of quantum information processing:
Alice and Bob cannot increase the total amount of entanglement that they share by local operations and classical communication alone.

[10]We will be more precise about what it means to quantify the amount of entanglement in a state shortly.

Contrary to the previous formulation, the phrase 'with no matter how small a probability' is missing. This law thus says that the total or, rather, expected entanglement cannot be increased. This still leaves room for the possibility that with some probability Alice and Bob can obtain a more entangled state. Then, however, with some other probability they will obtain less entangled states so that on average the entanglement will not increase. This law looks deceptively simple, but leads to some profound implications in quantum information processing. Although it is derived from considerations of the teleportation protocol, it nevertheless has much wider consequences. For example, we have established that if Alice and Bob share separable states of the form in Eq. (11.20) then no teleportation is possible. But what about the converse: if they share a state not of the form given in Eq. (11.20) can they always perform teleportation? More specifically, even if the state contains a small amount of entanglement, can that always be used for teleportation? This amounts to asking whether, given any entangled state, Alice and Bob can, with some probability, obtain the state $(|00\rangle + |11\rangle)/\sqrt{2}$ by acting only locally and communicating classically. This question will be addressed in the next section.

11.6. Entanglement Purification

In the previous section, we saw that starting from a product state and using only local operations and classical communication, the best we can achieve is a classically correlated state. We will never obtain a state that contains any quantum correlations and so we will not be able to teleport an unknown quantum state if we only share a classically correlated quantum state.

The impossibility of creating entanglement locally poses an important practical problem to Alice and Bob when they want to do teleportation in a realistic experimental situation. Imagine Alice wants to teleport a quantum state to Bob. Furthermore, assume that Alice and Bob are really far apart from each other and can exchange quantum states only for example through an optical fibre. The fibre, which we will occasionally call a quantum channel, is really long and it is inevitable that it contains faults, such as impurities, which will disturb the state of a photon that we send through the fibre. For teleportation Alice and Bob need to share a maximally entangled state, e.g. a singlet state. However, whenever Alice prepares a singlet state on her side and then sends one half of it to Bob the impurities in the fibre will disturb the singlet state. Therefore, after the transmission Alice and Bob will not share a singlet state but some mixed state that is no longer maximally entangled. If Alice attempts teleportation with this perturbed state, Bob will not receive the quantum state Alice tried to send but some perturbed (and usually mixed) state. Facing this situation Alice and Bob become quite desperate, because they have learnt that it is not possible to create quantum entanglement by local operations and classical communication alone. Because Alice and Bob are so far apart from each other, these are the only operations available to them. Therefore, Alice and Bob conclude that it will be impossible to 'repair' the state they are sharing in order to obtain a perfect singlet between them.

Luckily Alice and Bob have some friends who are physicists called say Charles, Gilles, Sandu, Benjamin, John and William[11] and they tell them of their predicament and ask for advice. In fact Charles, Gilles, Sandu, Benjamin, John and William confirm that it is impossible to create entanglement from nothing, i.e. local operations and classical communication starting with a product state. However, they inform Alice and Bob that while it is impossible to create quantum entanglement locally when you have no initial entanglement, you can in some sense amplify or concentrate entanglement from a source of weakly entangled states (this was the more general formulation of the fundamental law). We will now briefly explain two particular implementations (there are too many to discuss all of them) of these entanglement purification methods in order to show that they really work.

One main difference between the existing purification schemes is their generality, i.e. whether they can purify an arbitrary quantum state or just certain subclasses, such as pure states. Here we will present a scheme that can purify arbitrary (pure or mixed) bipartite states, if these states satisfy one general condition. This condition is expressed via the fidelity $F(\rho)$ of the state ρ, which is defined as

$$F(\rho) = \max_{\{all\; max.\; ent. |\psi\rangle\}} \langle\psi|\rho|\psi\rangle. \tag{11.21}$$

In this expression the maximization is taken over all maximally entangled states, i.e. over all states that one can obtain from a singlet state by local unitary operations. The scheme we are presenting here requires that the fidelity of the quantum state is larger than 0.5 in order for it to be purifiable.

Although it is possible to perform entanglement purification acting on a single pair of particles only, it can be shown that there are states that cannot be purified in this way[12]. Therefore, we present a scheme that acts on two pairs simultaneously. This means that Alice and Bob need to create initially two non-maximally entangled pairs of states which they then store. Now that Alice and Bob are holding the two pairs, they each perform two operations. First Alice performs a rotation on the two particles she is holding. This rotation has the effect that

$$|0\rangle \rightarrow \frac{|0\rangle - i|1\rangle}{\sqrt{2}} \tag{11.22}$$

$$|1\rangle \rightarrow \frac{|1\rangle - i|0\rangle}{\sqrt{2}}. \tag{11.23}$$

---

[11] These names, of course, have not been chosen randomly but are the names of the physicists who pioneered the idea of entanglement purification: Charles Bennett, Gilles Brassard, Sandu Popescu, Benjamin Schumacher, John Smolin and William Wootters.

[12] For more details, the reader is directed to the original journal article: Linden N., Massar S., Popescu S. (1998). Purifying Noisy Entanglement Requires Collective Measurements. *Physical Review Letters* **81**: 3279–3282 [14].

Bob performs the inverse of this operation on his particles, i.e.

$$|0\rangle \rightarrow \frac{|0\rangle + i|1\rangle}{\sqrt{2}} \tag{11.24}$$

$$|1\rangle \rightarrow \frac{|1\rangle + i|0\rangle}{\sqrt{2}}. \tag{11.25}$$

Subsequently both Alice and Bob perform a controlled NOT (CNOT) gate between the two particles they are holding. The effect of a CNOT gate is that the second bit gets inverted (NOT) when the first bit is in the state 1 while it remains unaffected when the first bit is in the state 0, i.e.

$$|0\rangle|0\rangle \rightarrow |0\rangle|0\rangle \tag{11.26}$$

$$|0\rangle|1\rangle \rightarrow |0\rangle|1\rangle \tag{11.27}$$

$$|1\rangle|0\rangle \rightarrow |1\rangle|1\rangle \tag{11.28}$$

$$|1\rangle|1\rangle \rightarrow |1\rangle|0\rangle. \tag{11.29}$$

The last step in the purification procedure consists of a measurement that both Alice and Bob perform on their particle of the second pair. They inform each other about the measurement result and keep the first pair if their results coincide. Otherwise they discard both pairs. In each step they, therefore, discard at least half of the pairs. From now on we are only interested in those pairs that are not discarded. In the Bell basis of Eqs. (11.3)-(11.6) we define the coefficients

$$A = \langle\Phi^+|\rho|\Phi^+\rangle \tag{11.30}$$

$$B = \langle\Psi^-|\rho|\Psi^-\rangle \tag{11.31}$$

$$C = \langle\Psi^+|\rho|\Psi^+\rangle \tag{11.32}$$

$$D = \langle\Phi^-|\rho|\Phi^-\rangle. \tag{11.33}$$

For the state of those pairs that they keep we find that

$$\tilde{A} = \frac{A^2 + B^2}{N} \tag{11.34}$$

$$\tilde{B} = \frac{2CD}{N} \tag{11.35}$$

$$\tilde{C} = \frac{C^2 + D^2}{N} \tag{11.36}$$

$$\tilde{D} = \frac{2AB}{N}. \tag{11.37}$$

Here $N = (A + B)^2 + (C + D)^2$ is the probability that Alice and Bob obtain the same results in their respective measurements of the second pair, i.e. the probability that they keep the first pair of particles. One can quite easily check that $\{A, B, C, D\} = \{1, 0, 0, 0\}$ is a fixed point of the mapping given in Eqs. (11.34 - 11.37) and that for $A > 0.5$ one also has $\tilde{A} > 0.5$. The ambitious reader might want to convince himself numerically that the fixed point $\{A, B, C, D\} = \{1, 0, 0, 0\}$ is indeed an attractor for all $A > 0.5$. The analytical proof of this is quite tricky. This means that if we initially have $A > 0.5$, our procedure will converge[13] towards the maximally entangled state $|\Phi^+\rangle$.

Now let us return to the problem of achieving teleportation over a noisy quantum channel and summarise what Alice and Bob have to do to achieve their goal. Initially they are given a quantum channel (for example an optical fibre) over which they can transmit quantum states. As this quantum channel is not perfect, Alice and Bob will end up with a partially entangled state after a single use of the fibre. Therefore, they repeat the transmission many times which gives them many partially entangled pairs of particles. Now they apply a purification procedure, such as the one described above, which will give them a smaller number of now maximally entangled pairs of particles. With these maximally entangled particles Alice and Bob can now teleport an unknown quantum state, e.g. $|\psi\rangle$ from Alice to Bob. Therefore, Alice and Bob can achieve perfect transmission of an unknown quantum state over a noisy quantum channel.

11.7. Purification of Pure States

The above title is an unfortunate choice of wording, because it might give the impression that we are trying to purify something that is already pure. However, you should remember that purification means entanglement concentration and pure states need not be maximally entangled. For example a pure state of the form $a|00\rangle + b|11\rangle$ is not maximally entangled unless $|a| = |b| = 1/\sqrt{2}$.

Let us see how we can purify pure states of this form by considering the following example. Alice and Bob share n entangled qubit pairs, where each pair is prepared in the state

$$|\Psi_{AB}\rangle = a|0_A 0_B\rangle + b|1_A 1_B\rangle, \tag{11.38}$$

where $|a|^2 + |b|^2 = 1$. How many maximally entangled states can they purify? It turns out that the answer is governed by the von Neumann reduced entropy[14] $S_{vN}(\rho_A) \equiv -\text{Tr}\{\rho_A \log_2 \rho_A\}$, where ρ_A is called the reduced density matrix and

[13]The map given by Eqs. (11.34)–(11.37) actually has two fixed points, namely $\{A, B, C, D\} = \{1, 0, 0, 0\}$ and $\{A, B, C, D\} = \{0, 0, 1, 0\}$. This means that if we want to know which maximally entangled state the procedure will converge towards, we need to have some more information about the initial state than just the fidelity given by Eq. (11.21).

[14]Entropy is usually defined with natural logarithms, i.e. logarithms with base e. However, in quantum information, the logarithms are usually taken to base 2. The reason for this is that it conveniently rescales things so that for a maximally entangled pair, such as one of the Bell states, the reduced entropy (and hence entanglement) has a value of one. We can see this by taking a particular example. Consider the Bell state $|\Phi^+\rangle$. The reduced density matrix for this is $\rho_A = (|0\rangle\langle 0| + |1\rangle\langle 1|)/2$, and so the entropy is

$$S_{vN} = -\frac{1}{2}\log_2\frac{1}{2} - \frac{1}{2}\log_2\frac{1}{2} = \log_2 2 = 1.$$

Of course, nothing changes physically when we pick different bases for the logarithms – we just rescale the units of our problem.

is obtained from the full density matrix, ρ, by tracing over subsystem B. For the example considered here (11.38), the density matrix is

$$\rho = |a|^2|0_A0_B\rangle\langle 0_A0_B| + |b|^2|1_A1_B\rangle\langle 1_A1_B| \\ +ab^*|0_A0_B\rangle\langle 1_A1_B| + a^*b|1_A1_B\rangle\langle 0_A0_B|, \tag{11.39}$$

and so the reduced density matrix ρ_A is

$$\rho_A = \mathrm{Tr}_B\{\rho\} = \langle 0_B|\rho|0_B\rangle + \langle 1_B|\rho|1_B\rangle \\ = |a|^2|0_A\rangle\langle 0_A| + |b|^2|1_A\rangle\langle 1_A|. \tag{11.40}$$

The total von Neumann reduced entropy for n copies of the state (11.38) is, therefore, $n \times S_{vN}(\rho_A) = n \times (-|a|^2 \log_2 |a|^2 - |b|^2 \log_2 |b|^2)$. It turns out that in the limit of large n this quantity is a measure of the number of maximally entangled states that can be purified from the total state. It is also important to stress that the above procedure is *reversible*. In other words, if Alice and Bob start by sharing $nS_{vN}(\rho_A)$ singlets then they can, at least in principle, locally produce n copies of the state $a|0,0\rangle + b|1,1\rangle$.

11.8. Entanglement Measures

We have seen that it is possible to concentrate entanglement using local operations and classical communication. A natural question that arises is how efficiently this can be done. To put it another way: given N partially entangled pairs of particles each in the state σ, how many maximally entangled pairs can we get? This question is basically about the amount of entanglement in a given quantum state. The more entanglement we have initially, the more singlet states we will be able to obtain from our supply of non-maximally entangled states. Of course one could also ask a different question, such as: how much entanglement do we need to create a given quantum state by local operations and classical communication alone? This question is somehow the inverse of the first one.

The answer to these questions lies in entanglement measures and in this section we will discuss these in a bit more detail. We will first explain the conditions that every 'reasonable' measure of entanglement should satisfy and then we will consider a particular example of an entanglement measure that is applicable to pure states.

To determine the basic properties any reasonable candidate for an entanglement measure should satisfy we need to look back on what we have discussed earlier in the chapter. We have seen that any state of the form Eq. (11.20), which we call separable, does not have any quantum correlations. This gives rise to our first condition:

1. For any separable state σ the measure of entanglement should be zero, i.e.

$$E(\sigma) = 0. \tag{11.41}$$

The next condition concerns the behaviour of the entanglement under simple local transformations, i.e. local unitary transformations. A local unitary transformation simply represents a change of the basis in which we consider the given entangled state. But a change of basis should not change the amount of entanglement that is accessible to us, because at any time we could just reverse the basis change. Therefore in both bases the entanglement should be the same.

2. For any state σ and any local unitary transformation, i.e. a unitary transformation of the form $U_A \otimes U_B$, the entanglement remains unchanged. Therefore,
$$E(\sigma) = E(U_A \otimes U_B \sigma U_A^\dagger \otimes U_B^\dagger). \tag{11.42}$$

The third condition is the one that really restricts the class of possible entanglement measures. Unfortunately it is usually also the property that is the most difficult to prove for potential measures of entanglement. We have already seen that Alice and Bob cannot create entanglement using only local operations and classical communication. We have also seen that, given some initial entanglement, we are able to select a subensemble of states that have higher entanglement. This can be done using only local operations and classical communication. However, what we cannot do is to increase the total amount of entanglement. We can calculate the total amount of entanglement by summing up the entanglement of all systems after we have applied our local operations, classical communications and subselection. We take the probability p_i that a system will be in particular subensemble $\mathcal{E}_i$ and multiply it by the average entanglement of that subensemble. We then sum up this result over all possible subensembles. The number we obtain should be smaller than the entanglement of the original ensemble.

3. Local operations, classical communication and subselection cannot increase the expected entanglement, i.e. if we start with an ensemble in state σ and end up with probability p_i in subensembles in state σ_i then we will have
$$E(\sigma) \geq \sum_i p_i E(\sigma_i). \tag{11.43}$$

This last condition has an important implication as it tells us something about the efficiency of the most general entanglement purification method. To see this we need to find out what the most efficient purification procedure will look like. Certainly it will select one subensemble, which is described by a maximally entangled state. For arguments sake, let us say this is the singlet state. As we want to make sure that we have as many pairs as possible in this subensemble, we assume that the entanglement in all the other subensembles vanishes. Then the probability that we obtain a maximally entangled singlet state from our optimal quantum state purification procedure is bounded by[15]
$$p_{\text{singlet}} \leq \frac{E(\sigma)}{E_{\text{singlet}}} = E(\sigma). \tag{11.44}$$

[15]The last step follows because our definition of von Neumann entropy is such that a singlet state has one unit of entanglement.

The considerations leading to Eq. (11.44) show that every entanglement measure that satisfies the three conditions presented in this section can be used to bound the efficiency of entanglement purification procedures from above. However, we should be careful as there is a hidden assumption in this argument, which is not quite trivial. We have assumed that the entanglement measures have the property that the entanglement of two pairs of particles is just the sum of the entanglements of the individual pairs. This sounds like a reasonable assumption but we should note that the entanglement measures that we construct are initially purely mathematical objects and that we need to prove that they behave reasonably. Therefore, we demand this additivity property as a fourth condition.

4. Given two pairs of entangled particles in the total state $\sigma = \sigma_1 \otimes \sigma_2$ then we have
$$E(\sigma) = E(\sigma_1) + E(\sigma_2). \tag{11.45}$$

Now that we have specified reasonable conditions that any measure of entanglement should satisfy, let us illustrate them by considering a particular example: the entropy of entanglement. Although this measure is defined only for pure states, it is nonetheless of great importance because there are good reasons to accept it as the unique measure of entanglement for pure states.

The entropy of entanglement is defined in the following way. Assume that Alice and Bob share an entangled pair of particles in a state σ. Then if Bob considers his particle alone he holds a particle whose state is described by the reduced density operator $\sigma_B = \mathrm{Tr}_A\{\sigma\}$. The entropy of entanglement is then defined as the von Neumann entropy of the reduced density operator σ_B, i.e.

$$E_{vN} = S_{vN}(\sigma_B) = -\mathrm{Tr}\{\sigma_B \log_2 \sigma_B\}. \tag{11.46}$$

One could think that the definition of the entropy of entanglement depends on whether Alice or Bob calculate the entropy of their reduced density operator. However, it can be shown that for a pure state σ this is not the case, i.e. both will find the same result. It can be shown that this measure of entanglement, when applied to pure states, satisfies all the conditions that we have formulated in the previous section. This certainly makes it a good measure of entanglement. In fact many people believe that it is the only measure of entanglement for pure states. Why is that so? In the previous section we have learnt that an entanglement measure provides an upper bound to the efficiency of any purification procedure. For pure states it has been shown that there is a purification procedure that achieves the limit given by the entropy of entanglement. In addition the inverse property has also been shown. Assume that we want to create N copies of a quantum state σ of two particles purely by local operations and classical communication. As local operations cannot create entanglement, it will usually be necessary for Alice and Bob to share some singlets before they can create the state σ. How many singlet states do they have to share beforehand? The answer, again, is given by the entropy of entanglement, i.e. to create N copies of a state σ of two-particles one needs to share $N\,E(\sigma)$ singlet states beforehand. Therefore, we have a very interesting result. The entanglement of pure states can be concentrated and subsequently be diluted again in a reversible fashion. One should note, however, that this result holds only when we have many (actually infinitely many) copies of entangled pairs at our disposal. For finite N it is not possible to achieve the theoretical limit exactly. This observation suggests a close relationship between entanglement transformations of pure states and thermodynamics[16]. In the next section we will elaborate on this connection further.

11.9. Thermodynamics of Entanglement

Here we would like to elucidate further the fundamental law of quantum information processing by comparing it to the second law of thermodynamics. The reader should not be surprised that there are connections between the two. First of all, both laws can be expressed mathematically by using an entropic quantity. The second law says that thermodynamical entropy cannot decrease in an isolated system. The fundamental law of quantum information processing, on the other hand, says that entanglement cannot be increased by local operations. Thus both of the laws serve to prohibit certain types of processes, which are impossible in nature.

We begin by stating more formally a form of the second law of thermodynamics. This form is due to Clausius, but it is completely analogous to the no increase of entropy statement we gave above. In particular it will be more useful for what we are about to investigate.

The Second Law of thermodynamics (Clausius):
There exists no thermodynamic process the *sole* effect of which is to extract a quantity of heat from the colder of two reservoirs and deliver it to the hotter of the two reservoirs.

Suppose now that we have a thermodynamical system. We want to invest some heat Q_{in} into it so that at the end our system does as much work, W_{out}, as possible with this heat input. The efficiency is, therefore, defined as

$$\eta = \frac{W_{\text{out}}}{Q_{\text{in}}}. \tag{11.47}$$

Now it is a well known fact that the above efficiency is maximised if we have a reversible process simply because an irreversible process wastes useful work on friction or some other lossy mechanism. In fact, we know the efficiency of one such process, called the Carnot cycle. With the second law in mind, we can now prove that no other process can perform better than the Carnot cycle. This boils down to the fact that we only need to prove that no other reversible process

[16]It is also possible to generalise the entropy of entanglement to mixed states. However, for mixed states there is not one unique measure of entanglement but several different ones. This is a very interesting topic in its own right.

performs better than the Carnot cycle. The argument for this can be found in any undergraduate book on thermodynamics and briefly runs as follows (again using *reductio ad absurdum*). The Carnot engine takes some heat input from a hotter reservoir, does some work and delivers an amount of heat to the colder reservoir. Suppose that there is a better engine E that is operating between the same two reservoirs (we have to be fair when comparing the efficiency). Suppose also that we run this better machine backwards as a refrigerator. We would do some work on it and it would take a quantity of heat from the cold reservoir and bring some heat to the hot reservoir. For simplicity we assume that the work done by a Carnot engine is the same as the work that E needs to run in reverse (this can always be arranged and we lose nothing in generality). Then we look at the two machines together, which is just another thermodynamical process. They extract a quantity of heat from the colder reservoir and deliver it to the hot reservoir with all other things being equal. However this contradicts the second law and, therefore, no machine is more efficient than the Carnot engine.

In the previous section we discussed a purification (or entanglement concentration) scheme for pure states. Let us call this P. The efficiency of such a scheme is defined as the number of maximally entangled states we can obtain from a given N pairs in some initial state, divided by N. This scheme is in addition reversible and we would suppose, guided by the above thermodynamic argument, that no other reversible purification scheme could do better than P. Suppose that there is a more efficient (reversible) process. Now Alice and Bob start with a certain number N of maximally entangled pairs. They apply a reverse of the scheme of P to get a certain number of less entangled states. But then they can run the more efficient purification to get M maximally entangled states out. However, since the second purification is more efficient than the first one, then we have that $M > N$. So, locally Alice and Bob can increase entanglement, which contradicts the fundamental law of quantum information processing.

This shows that the conceptual ideas behind the second law and the fundamental law are similar in nature. Next we show another attractive application of the fundamental law. We return to the question we raised earlier: can Alice teleport to Bob as many qubits as she likes using only one entangled pair shared between them?

It would certainly be very useful if a single maximally entangled pair could transfer a large amount of information by being used to repeatedly teleport different qubits. However, there is no such thing as a free lunch. In the same way that we cannot have an unlimited amount of useful work and no heat dissipation, we cannot have arbitrarily many teleportations with a single maximally entangled pair. In fact, we can prove a much stronger statement. In order to teleport N qubits Alice and Bob need to share N maximally entangled pairs.

In order to prove this we need to understand another simple concept from quantum mechanics. Namely, if we can teleport a pure unknown quantum state then we can teleport an unknown mixed quantum state (this is obvious since a mixed state is just a combination of pure states). But now comes a crucial result: every mixed state of a single qubit can be thought of as a part of a pure state of two *entangled* qubits (this result is more general, and applies to any quantum state of any quantum system, but we do not need the generalisation here). For example, suppose that we have a single qubit in a state

$$\rho = a^2|0\rangle\langle 0| + b^2|1\rangle\langle 1| . \tag{11.48}$$

This single qubit can then be viewed as a part of a pair of qubits in state

$$|\psi\rangle = a|00\rangle + b|11\rangle . \tag{11.49}$$

One obtains Eq. (11.48) from Eq. (11.49) simply by taking the partial trace over the second particle, as we saw in Eq. (11.40). Bearing this in mind we now envisage the following teleportation protocol. Alice and Bob share a maximally entangled pair, and in addition Bob has a qubit prepared in some state, say $|0\rangle$. Alice then receives a qubit to teleport in a general state ρ. After the teleportation we want Bob's extra qubit to be in the state ρ and the maximally entangled pair to stay intact or, at least, not to be completely destroyed.

Now we wish to prove that this protocol is impossible – entanglement simply has to be completely destroyed at the end. Suppose it is not, i.e. suppose that the above teleportation is possible. Then Alice can teleport any unknown (mixed) state to Bob using this protocol. But this mixed state can arise from an entangled state where the second qubit (the one to be traced out) is on Alice's side. So initially Alice and Bob share one entangled pair, but after the teleportation they have increased their entanglement. Since the initial state can be a maximally mixed state ($a = b = 1/\sqrt{2}$) the final entanglement can grow to be twice the maximally entangled state. But, as this would violate the fundamental law of quantum information processing it is impossible and the initial maximally entangled pair has to be destroyed. In fact, this argument shows that it has to be destroyed completely. Thus we see that a simple application of the fundamental law can be used to rule out a whole class of impossible teleportation protocols. Otherwise every teleportation protocol would have to be checked separately and this would be a very hard problem.

11.10. Quantum Computing

Entanglement has many potential technological applications. One of the most exciting of these is the quantum computer. In ordinary 'classical' computing information is stored as bits that can take the value of 0 or 1. These bits are processed by passing them through logical gates to perform calculations. In practice, this all takes place on a microchip with electrons acting as the bits and transistors playing the role of the gates.

Classical computing has, of course, been a runaway success. Not only has it spawned a vast industry employing many millions of people, but it has revolutionised society and our lives. One of the key drivers of the success of this industry has been the rate at which improvements in computing power have been able to be achieved. This rate of progress is captured in a concept known as Moore's law that was put forward by Gordon Moore (one of the founders of Intel) in 1965. He said that the number of transistors that can be placed inexpensively on an integrated circuit doubles every two years[17]. In other words, computers should get twice as fast every two years without significantly increasing in cost. This law has held remarkably accurately for the past 40 or so years.

The problem is that Moore's law cannot go on indefinitely. Moore himself recently said: "Any physical quantity that's growing exponentially predicts a disaster. You simply can't go beyond certain major limits." One of the main problems is that circuit dimensions are now on the scale of nanometers and cannot get much smaller without fundamental alterations in how semiconductors are made. There are also major heating problems with circuits on such small scales. It looks likely, therefore, that Moore's law will soon run out of steam. Does this means that the advances in computing power that we have become so accustomed to will grind to a halt?

The answer is hopefully not. One possible way round the problem is offered by quantum computing – a whole new concept in computing which exploits the power of entanglement. While ordinary computers use bits that are in one state or the other, quantum computers use qubits which, as we have seen, can be in any arbitrary superposition of 0 and 1. Different qubits can also be entangled with one another. This offers the possibility of massive parallelism since, in quantum mechanics, n quantum systems can represent 2^n numbers simultaneously. However, building a quantum computer is a tricky business. The disruptive influence of the environment makes the realisation of quantum computing extremely difficult and many ideas have been developed to combat the noise – incidentally again using entanglement.

Once realised, a quantum computer could offer an exponential increase of computational speed for certain problems, such as the factorisation of large numbers into primes. This is an important problem that is at the very heart of electronic encryption schemes and relies on the fact that it is very hard for classical computers to factorise large numbers, but very easy to check whether a given solution is correct. This suggests that a quantum computer could undermine internet security in one fell swoop. Fortunately, there is another branch of research called quantum cryptography that is dedicated to creating ways of distributing messages in a way that can never be eavesdropped upon. Many of these schemes rely on, once again, entanglement to work. It is even possible to argue that the whole universe can be thought of as a quantum computer[18]. These and many other applications of entanglement form the subject of a very interesting and active field of research called quantum information[19].

---

[17]Actually he first said that the number doubles every year, but later refined this period to two years. Confusingly, it is often quoted as a doubling of the number of transistors every 18 months.

[18]For a popular account of this idea, see *Decoding Reality: The Universe as Quantum Information* by V. Vedral (Oxford, 2010) [15].

[19]For much more information about this field, we direct the reader to the book *Quantum Computation and Quantum Information* by Nielsen and Chuang (Cambridge, 2000) [11] or *Introduction to Quantum Information Science* by V. Vedral (Oxford, 2006) [12].

Quantum information theory is still a relatively new field and, although there are already some extraordinary results, there is still a lot to be done. It will be interesting, for example, to develop a deeper understanding of the relationship between information theory, quantum physics and thermodynamics. Quantum theory has had a huge input into information theory and thermodynamics over the past few decades. Perhaps by turning this around we can learn much more about quantum theory by using information-theoretic and thermodynamic concepts. Ultimately, this approach might solve some long standing and difficult problems in modern physics, such as the measurement problem and the arrow of time problem. This is exactly what was envisaged more that 60 years ago in a statement attributed to Einstein: "The solution of the problems of quantum mechanics will be thermodynamical in nature".

11.11. Outlook

Now that we have come to the end of the book it is worth pausing to review what we have covered. Our journey started in the late nineteenth century. This was a time when confidence in the success of physics was sky-high, but there were still some niggling experimental results that couldn't be explained. In April 1900, Lord Kelvin gave a lecture to the Royal Institution in which he said that the 'beauty and clearness' of physical theories were overshadowed by 'two clouds' (what David Hume would call 'black swans'). He was talking about the null result of the Michelson-Morley experiment and the problems of black body radiation. We now know that these 'two clouds' were precisely what led to the emergence of relativity and quantum theory and the advent of what came to be known as the "Century of Physics".

In this book, we have traced the development of these theories. Wherever possible we have aimed to do this historically and to present arguments in ways similar to how they were first proposed. This, we hope, has helped give a flavour of how science works. Of course the picture we have painted is biased. Space limitations mean that we have not recorded the many failures along the way and even some of the successes that we have reported glibbly in a few lines may represent the outcome of many years of toil.

We started out by showing how the fathers of quantum physics used analogies with thermodynamics to argue that the world really must come in discrete chunks or quanta. This was a strange and controversial claim at the time, but it led to the formal theory of quantum mechanics, which was able to explain experimental results beautifully. At about the same time, Einstein was breathing new life into the mathematical transformations developed by Lorentz and FitzGerald. His theory of relativity forever changed our ideas of space, time and simultaneity. Remarkably, these two theories have been able to be combined by the likes of Dirac, Klein and Gordon to give us relativistic quantum mechanics. This in turn led to quantum field theory – a true intellectual triumph and the most precise scientific theory that we have. Finally, we have seen how quantum field theory led us to the intriguing ideas of entanglement and quantum information and its connections back to thermodynamics – right where we began. This interconnectedness is part of the great beauty of science and is what helps us to spot patterns and relationships that allow us to push the boundaries further. Our discussion has taken us to the forefront of current research and highlighted some of the questions facing current researchers. Along the way, we have discussed some of the important applications of quantum physics: from scanning tunnelling microscopes to lasers and quantum computers.

As we said at the beginning, one of our main aims has been to convey how quantum physics and relativity form an important part of our cultural and intellectual heritage. We hope you have also come to appreciate the beauty of the scientific method itself and the remarkable results it can achieve. In some ways science is perverse. It can only prove things wrong, but we shouldn't shy away from Hume's black swans. Contradictory evidence is one of most exciting things that can happen in physics, because it often means we are on the cusp of unveiling a new and better theory[20]. There is still a lot of work to be done and a number of issues are still unresolved. We can be sure that the theories we have presented in this book will someday be surpassed by superior, more general ones – that is the excitement of science. What the future holds, only time can tell.

[20] As Niels Bohr once said: "How wonderful we have met with a paradox, now we have some hope of making progress".

11.12. Exercises

1. Calculate the entropy for a quantum state with the density matrix

$$\rho = \begin{pmatrix} 1/2 & 1/4 \\ 1/4 & 1/2 \end{pmatrix}.$$

2. Calculate the entropy of entanglement for the state $a|00\rangle + b|11\rangle$. For what values of a and b is the entanglement highest? When is it zero? Explain your results.

3. (a) Suppose you have an atom that has been prepared in an equal mixture of the ground state, $|g\rangle$, the excited state, $|e\rangle$, and an equal superposition of the ground and excited state, i.e. $(|g\rangle + |e\rangle)/\sqrt{2}$. Write down the density matrix that represents the state of the atom. What is the entropy of this state?

   (b) If you subsequently measure the atom in the basis of the ground and excited states, what is the resulting density matrix describing your final state (assuming that you do not observe the actual outcome)?

   (c) Is the entropy after the measurement smaller or bigger than before? Comment on the result.

4. (a) An experimentalist trying to teleport is unable to achieve a maximally entangled state with perfect efficiency. There is always a certain amount of noise involved. Suppose that the state $|\Phi^+\rangle = (|00\rangle + |11\rangle)/\sqrt{2}$ can be created with 90% efficiency and the remaining 10% belongs to the state $|00\rangle$. Write down the corresponding density matrix of the state.

   (b) Now use this state in the teleportation protocol. What is the state of the output (teleported) qubit?

Chapter 12

Solutions

12.1. Chapter 2

1. From the given relationship

$$\frac{p^2}{2M} = \frac{3kT}{2}$$

we conclude, after rearrangement, that the thermal momentum is

$$p = \sqrt{3MkT}.$$

Quantum effects become important when the de Broglie wavelength of individual systems becomes comparable to the distance between them. Thus we have that

$$\lambda = \frac{h}{p} = d.$$

From this we can conclude that for quantum effects to be important the momentum should obey

$$p = \frac{h}{d}.$$

Equating this value of the momentum to the thermal momentum and re-expressing for temperature leads us to the following equation:

$$T = \frac{h^2}{3Mkd^2}.$$

Putting in the numbers for electrons (M_e is needed)

$$T = \frac{h^2}{3M_e kd^2} \approx 10^5\text{K},$$

while for nuclei (M_n is needed)

$$T = \frac{h^2}{3M_n k d^2} \approx 70\text{K}.$$

Therefore at room temperature ($T = 300$K), electrons in a solid are quantum (since their critical temperature is way above this), while nuclei can be assumed to be classical (since for them to be quantum they need to be cooled below 70K).

2. The average distance between atoms is

$$d = \frac{N}{L} = \frac{1}{\rho} = \frac{h}{\sqrt{3MkT}}.$$

Rearranging for temperature we obtain

$$T = \frac{h^2}{3Mk}\rho^2.$$

If $L = 100\mu$m and $N = 100000$ we have

$$T \approx 0.1\text{K}.$$

Therefore, in one dimension atoms would condense below this temperature. In three dimensions, on the other hand,

$$d = \left(\frac{V}{N}\right)^{1/3} = \left(\frac{1}{\rho}\right)^{1/3}.$$

Therefore, the condensation temperature is now

$$T = \frac{h^2}{3Mk}\rho^{2/3} \approx 10^{-8}\text{K}.$$

This is close to the temperatures observed in actual experiments.

3. The power emitted by the Sun is

$$P_S = \sigma T_S^4 \times 4\pi R_S^2.$$

This is just the Stefan–Boltzmann expression for intensity per unit time multiplied by the area of the Sun. The power absorbed by the Earth, assuming that it is a black body, is

$$P_E^{abs} = P_S \times \frac{\pi R_E^2}{4\pi D^2} = \sigma T_S^4 \times 4\pi R_S^2 \times \frac{\pi R_E^2}{4\pi D^2},$$

where D is the distance between the Earth and the Sun. This formula is obtained by multiplying the power of the Sun by the fraction of this power absorbed by the Earth.

The power emitted by the Earth is

$$P_E^{emit} = \sigma T_E^4 \times 4\pi R_E^2$$

in direct analogy with the power emitted by the Sun.

For equilibrium we need

$$P_E^{emit} = P_E^{abs}.$$

From this we obtain that

$$T_E = \sqrt{\frac{R_S}{2D}} T_S.$$

Plugging in the numbers ($D = 1.5 \times 10^{11}$m and $R_S = 7 \times 10^8$m and $T_S = 5780$), we get

$$T_E \approx 279\text{K}.$$

This is lower than the actual temperature. The main reason for higher T is the atmosphere, i.e. the greenhouse effect.

4. The net power for a human is

$$\sigma(T_H^4 - T_S^4)A = 95\text{W}.$$

This is almost the same as a 100W bulb.

The wavelength at which the power is maximal is

$$\lambda_m = \frac{3 \times 10^{-4}}{300K} = 10^{-6}\text{m} = 1000\,\text{nm}.$$

This is outside of the visible spectrum (300 – 700 nm). You, in fact, need infrared goggles to see this.

5. The formula for the wavelength of light emitted in the $n+1$ to n transition is

$$\frac{1}{\lambda} = R\left[\frac{1}{n^2} - \frac{1}{(n+1)^2}\right],$$

where $R = 10^7 m^{-1}$ is the Rydberg constant. The transition for $n = 1$ gives us

$$\lambda \approx 100\,\text{nm}$$

and this is, of course, outside the visible spectrum (ultraviolet). In general

$$\frac{1}{\lambda} = R\frac{(n+1)^2 - n^2}{n^2(n+1)^2} \approx R\frac{2}{n(n+1)^2}.$$

For $n = 3$ the corresponding wavelength is in the visible domain.

The appropriate mass correction can be obtained by imagining that both the electron as well as the nucleus orbit about the common centre of mass. The condition for centre of mass is that

$$M_e r_e = M_n r_n = \mu(r_e + r_n),$$

from which we can derive that the effective mass is

$$\mu = \frac{M_e M_n}{M_e + M_n}.$$

The effective mass is smaller than the actual one by only about one part in ten thousand.

$$\mu = 10^{-31} \frac{1}{1 + 10^{-4}} \text{kg}.$$

This is the same as the error introduced to the wavelength since

$$\Delta\lambda \propto R \propto M_e \to \mu.$$

This is the same as modifying the Rydberg constant by

$$R \to \frac{R}{1 + M_e/M_n}.$$

12.2. Chapter 3

1. The probability density is

$$|e^{i\omega t - kx} + e^{i\omega t + kx}|^2 \propto 1 + \cos(2kx).$$

Nodes are when density is zero, so

$$2kx = (2n - 1)\pi.$$

Maxima occur at

$$2kx = n\pi.$$

The nodes are minima of the probability density.

The current is identically zero in this case. Each plane wave has equal current

$$j = \frac{\hbar k}{M}.$$

but in with opposite sign (i.e. direction) and the sum is, therefore, zero. This makes sense since this wave is standing, which means it is not travelling, and its current, i.e. flow, therefore, has to vanish.

2. The normalisation integral is

$$A^2 \int_{-a}^{a} (a^2 - x^2)^2 dx = 1,$$

from which we can deduce that

$$A = \sqrt{\frac{15}{16a^5}}.$$

It is clear that averages of x and p are zero since they involve integrals of odd functions.

We now evaluate

$$\langle x^2 \rangle = \frac{15}{16a^5} \int_{-a}^{a} x^2 (a^2 - x^2)^2 dx = \frac{a^2}{7}.$$

As for the momentum we get

$$\langle p^2 \rangle = \frac{15\hbar^2}{16a^5} \int_{-a}^{a} (a^2 - x^2)(2) dx = \frac{5\hbar^2}{2a^2}.$$

The product of the uncertainties in position and momentum is, therefore,

$$\Delta x \Delta p = \sqrt{\frac{a^2}{7} \frac{5\hbar^2}{1a^2}} = 0.6\hbar > \frac{\hbar}{2}$$

and this confirms the Heisenberg uncertainty relations.

If a halves, then the uncertainty in momentum doubles because $\Delta p \propto 1/a$.

3. To show that probability is conserved it is sufficient to show that

$$\frac{d}{dt}(\psi^* \psi) = 0,$$

but this easily follows from the fact that

$$\psi^* \psi = e^{-i(\omega t - kx)} e^{i(\omega t - kx)} = 1.$$

We obtain $A(t)$ from

$$A^2 \int_0^L e^{-i(\omega t - kx)} e^{i(\omega t - kx)} dx = A^2 L = e^{-t/\tau},$$

therefore,

$$A = \frac{e^{-t/2\tau}}{\sqrt{L}}.$$

In order for

$$\Psi(x,t) = \frac{e^{-t/2\tau}}{\sqrt{L}} e^{-(\omega t - kx)}$$

to be the solution to the Schrödinger equation, we need

$$V = \frac{i\hbar}{2\tau}.$$

If the decay results in two photons the sum of their frequencies should match the energy of decay, i.e.

$$\frac{1}{2\tau} = 2\omega_p,$$

which means the wavelength of the photons is

$$\lambda = \frac{2\pi c}{\omega_p} = 8\pi c\tau.$$

4.

$$\begin{aligned}
-\frac{\hbar^2}{2M}\frac{\partial^2\Psi}{\partial x^2} + V(x,t)\Psi &= -\frac{\hbar^2}{2M}\frac{\partial^2}{\partial x^2}(c_1\psi_1 + c_2\psi_2) + V(x,t)(c_1\psi_1 + c_2\psi_2) \\
&= c_1\left[-\frac{\hbar^2}{2M}\frac{\partial^2\psi_1}{\partial x^2} + V(x,t)\psi_1\right] + c_2\left[-\frac{\hbar^2}{2M}\frac{\partial^2\psi_2}{\partial x^2} + V(x,t)\psi_2\right] \\
&= c_1\left[i\hbar\frac{\partial\psi_1}{\partial t}\right] + c_2\left[i\hbar\frac{\partial\psi_1}{\partial t}\right] \\
&= i\hbar\frac{\partial}{\partial t}(c_1\psi_1 + c_2\psi_2) = i\hbar\frac{\partial\Psi}{\partial t}.
\end{aligned}$$

This means we have

$$-\frac{\hbar^2}{2M}\frac{\partial^2\Psi}{\partial x^2} + V(x,t)\Psi = i\hbar\frac{\partial\Psi}{\partial t},$$

as required.

12.3. Chapter 4

1. By definition, this is

$$P_{1/3} = \frac{2}{L}\int_0^{L/3} \sin^2(\pi/Lx)dx.$$

After the change of variables $y = \pi/Lx$ we get

$$P_{1/3} = \frac{2}{\pi}\left[\frac{y}{2} - \frac{\sin 2y}{4}\right]_0^{\pi/3} = 0.2.$$

The probability to be in the middle third is equal to

$$P_{\text{middle}} = 1 - 2P_{1/3} = 0.6.$$

For high quantum numbers n

$$P_{1/3} = \frac{2}{L}\int_0^{L/3} \sin^2(n\pi/Lx)dx.$$

When $n \to \infty$

$$P_{1/3} \to \frac{2}{\pi}\left[\frac{y}{2}\right]_0^{\pi/3} = \frac{1}{3}$$

and this makes sense because the probability density for a classical particle is uniform everywhere.

2. Each state evolves according to its own energy eigenvalue

$$\Psi_n(x,t) = \sin\left(\frac{n\pi x}{L}\right)e^{-iE_n t/\hbar},$$

where

$$E_n = \frac{\pi^2\hbar^2}{2ML^2}n^2$$

Due to linearity of Schrödinger's equation we have that the state at t is

$$\Psi_1(x,t) + \Psi_2(x,t) \propto \sin\left(\frac{\pi x}{L}\right)e^{-iE_1 t/\hbar} + \sin\left(\frac{2\pi x}{L}\right)e^{-iE_2 t/\hbar}.$$

The probability to obtain the ground state at any time is the same as at the beginning and it is one half.
We obtain the same state as the initial one ($t = 0$) when

$$\left(\frac{\pi^2\hbar}{2ML^2}2^2 - \frac{\pi^2\hbar}{2ML^2}1^2\right)t = \frac{3\pi^2\hbar}{2ML^2}t = 2\pi$$

from which we obtain

$$t = \frac{4ML^2}{3\pi\hbar}.$$

We can compute the velocity of the corresponding classical particle from

$$\frac{Mv^2}{2} = E_2 - E_1 = \frac{3\pi^2\hbar^2}{2ML^2}$$

The time for the classical particle is given by

$$t_c = \frac{L}{v} = \frac{\sqrt{3}ML^2}{\sqrt{3}\pi\hbar},$$

so the classical time is very similar to the quantum one.

3. (c)

4. (a) See Fig. 4.2

(b) Equating the energy that a particle has is the ground state of an harmonic oscillator ($\hbar\omega/2$), to the potential energy it has at displacement $x = b$ from the equilibrium position, we get

$$\frac{1}{2}\hbar\omega = \frac{1}{2}M\omega^2 b^2.$$

Solving for b gives $b = \sqrt{\hbar/M\omega}$.

(c)

$$\begin{aligned} \frac{d\psi}{dx} &= -\frac{M\omega}{\hbar} x A \exp\left(\frac{-M\omega x^2}{2\hbar}\right) \\ \frac{d^2\psi}{dx^2} &= \left[-\frac{M\omega}{\hbar} + \frac{M^2\omega^2 x^2}{\hbar^2}\right] A \exp\left(\frac{-M\omega x^2}{2\hbar}\right), \end{aligned}$$

where the second line follows from the product rule. Substituting into the Schrödinger equation and dividing through by ψ, we get

$$\frac{-\hbar^2}{2M}\left[\frac{-M\omega}{\hbar} + \frac{M^2\omega^2 x^2}{\hbar^2}\right] = E - \frac{1}{2}M\omega^2 x^2.$$

This reduces to $E = \frac{1}{2}\hbar\omega$, i.e. $\psi = A\exp(\frac{1}{2}M\omega x^2/\hbar)$ is a solution for $E = \frac{1}{2}\hbar\omega$.

5. (a) We require

$$\int_{-\infty}^{\infty} dx\, \psi_1^*(x)\psi_1(x) = 1,$$

i.e.

$$|A|^2 \int_{-\infty}^{\infty} dx\, x^2 e^{-x^2/b^2} = 1.$$

Substituting $u = x/b$, we get,

$$b^3|A|^2 \int_{-\infty}^{\infty} du\, u^2\, e^{-u^2} = 1,$$

and using the given integral result, we get $b^3|A|^2\sqrt{\pi}/2 = 1$. Solving for A gives

$$A = \left(\frac{2}{\sqrt{\pi}b^3}\right)^{1/2}.$$

(b) Orthogonality requires

$$\int_{-\infty}^{\infty} \psi_1^*(x)\psi_0(x)\,dx = 0.$$

Substituting, we get

$$\begin{aligned}\left(\frac{2}{\sqrt{\pi}b^3}\right)^{1/2}\left(\frac{1}{\sqrt{\pi}b}\right)^{1/2}&\int_{-\infty}^{\infty} xe^{-x^2/2b^2}e^{-x^2/2b^2}\,dx\\ &= \frac{1}{b^2}\sqrt{\frac{2}{\pi}}\int_{-\infty}^{\infty} xe^{-x^2/b^2}\,dx = 0.\end{aligned}$$

Where the last integral is zero because the integrand is odd. This confirms that the two states are orthogonal.

(c) The derivative of $\psi_1(x)$ can be found using the product rule,

$$\begin{aligned}\frac{d\psi_1}{dx} &= \left(\frac{2}{\sqrt{\pi}b^3}\right)^{1/2}\left[1-\frac{x^2}{b^2}\right]e^{-x^2/2b^2}\\ \frac{d^2\psi_1}{dx^2} &= \left(\frac{2}{\sqrt{\pi}b^3}\right)^{1/2}\frac{1}{b^2}\left[-3+\frac{x^2}{b^2}\right]xe^{-x^2/2b^2} = \frac{1}{b^2}\left[-3+\frac{x^2}{b^2}\right]\psi_1(x)\end{aligned}$$

This means we have

$$\begin{aligned}H\psi_1(x) &= \frac{-\hbar^2}{2M}\frac{M\omega}{\hbar}\left[-3+x^2\frac{M\omega}{\hbar}\right]\psi_1(x)+\frac{1}{2}M\omega^2x^2\psi_1(x)\\ &= \frac{3}{2}\hbar\omega\,\psi_1(x).\end{aligned}$$

Hence ψ_1 is an eigenstate of H with eigenvalue $\frac{3}{2}\hbar\omega$.

(d) For $x < 0$, $\psi = 0$ and for $x \geq 0$, ψ is given by the harmonic oscillator wave functions. The only harmonic oscillator wave functions that satisfy the boundary conditions, i.e. $\psi = 0$ at $x = 0$, are the odd ones. Therefore, the energy levels are: $E = \frac{3}{2}\hbar\omega, \frac{7}{2}\hbar\omega, \frac{11}{2}\hbar\omega, \ldots$

6. The probability amplitude that the particle is in the ground state of the expanded well is simply the overlap of the ground state of the expanded well, $|\psi_2\rangle$, with the ground state of the original well, $|\psi_2\rangle$, i.e. $\langle\psi_2|\psi_1\rangle$. Substituting,

$$\begin{aligned}\psi_1 &= \sqrt{\frac{2}{L}}\sin\left(\frac{\pi x}{L}\right)\\ \psi_2 &= \sqrt{\frac{1}{L}}\sin\left(\frac{\pi x}{2L}\right),\end{aligned}$$

we get,

$$\begin{aligned}\langle\psi_2|\psi_1\rangle &= \frac{\sqrt{2}}{L}\int_0^L \sin\left(\frac{\pi x}{L}\right)\sin\left(\frac{\pi x}{2L}\right)\,dx\\ &= \frac{1}{\sqrt{2}L}\int_0^L\left[\cos\left(\frac{\pi x}{2L}\right)-\cos\left(\frac{3\pi x}{2L}\right)\right]dx\\ &= \left[\frac{\sqrt{2}}{\pi}\sin\left(\frac{\pi x}{2L}\right)\right]_0^L-\left[\frac{\sqrt{2}}{3\pi}\sin\left(\frac{3\pi x}{2L}\right)\right]_0^L\\ &= \frac{\sqrt{2}}{\pi}+\frac{\sqrt{2}}{3\pi}=\frac{4\sqrt{2}}{3\pi}.\end{aligned}$$

So the probability that the particle is in the ground state of the expanded potential is

$$P_0 = \frac{32}{9\pi^2} \approx 0.36.$$

Using the same technique, the probabilities that the particle is found in the first and second excited states of the expanded potential are respectively,

$$\begin{aligned}P_1 &= 0.5\\ P_2 &= \frac{32}{25\pi^2} \approx 0.13.\end{aligned}$$

Adding these up, we get $P_0 + P_1 + P_2 \approx 0.99$, i.e. the particle will almost certainly be found in one of the three lowest levels of the expanded potential.

7. The transmission probability can be calculated using Eq. (4.32), i.e.

$$T(E) \approx \exp\left(-\frac{Ze^2}{\epsilon_0\hbar}\sqrt{\frac{M_\alpha}{2E}}+\frac{4e}{\hbar}\sqrt{\frac{ZRM_\alpha}{\pi\epsilon_0}}\right).$$

Substituting values: $Z = 88$ (daughter nucleus), $\epsilon_0 = 8.85\times10^{-12}$ Fm^{-1}, $M_\alpha = 4\times1.67\times10^{-27}$kg, $R = 9\times10^{-15}$m, and $E = 4.05$MeV, we get $T(E) \approx 1.29\times10^{-39}$. If we then assume that there are 10^{21} collisions per second, the decay rate is $\lambda = 10^{21}T(E) \approx 1.29\times10^{-18}$. Finally, the half-life is given by

$$t_{1/2} = \frac{\ln 2}{\lambda} = \frac{\ln 2}{1.29\times10^{-18}} \approx 5.37\times10^{17}\text{ s} \approx 1.7\times10^{10}\text{ years}.$$

This agrees well with the observed value (1.3×10^{10} years) given the crude nature of the model.

12.4. Chapter 5

1.

$$-i\frac{d\Phi(\phi)}{d\phi} = m\Phi(\phi)$$
$$\Longrightarrow \Phi(\phi) = \Phi(0)\exp(im\phi).$$

Using the fact that this function should not change under the transformation $\phi \to \phi + 2\pi$, we see that m must be an integer. Now, identifying,

$$L_z = -i\hbar\frac{d}{d\phi},$$

we get

$$L_z\Phi(\phi) = \hbar m\Phi(\phi).$$

So $m\hbar$ is the z-component of angular momentum.

2. Multiplying both sides of Eq. (5.14) on the right by $\Phi(\phi)$ we get

$$\left[-\frac{1}{\sin\theta}\frac{\partial}{\partial\theta}\left(\sin\theta\frac{\partial}{\partial\theta}\right) + \frac{m^2}{\sin^2\theta}\right]\Theta(\theta)\Phi(\phi) = l(l+1)\Theta(\theta)\Phi(\phi).$$

We then use Eq. (5.13) to observe

$$m^2\Theta(\theta)\Phi(\phi) = \left[-i\frac{\partial}{\partial\phi}\right]^2\Theta(\theta)\Phi(\phi) = -\frac{\partial}{\partial\phi}\Theta(\theta)\Phi(\phi).$$

Substituting into the first equation, we get

$$\left[-\frac{1}{\sin\theta}\frac{\partial}{\partial\theta}\left(\sin\theta\frac{\partial}{\partial\theta}\right) - \frac{1}{\sin^2\theta}\frac{\partial}{\partial\phi}\right]\Theta(\theta)\Phi(\phi) = l(l+1)\Theta(\theta)\Phi(\phi).$$

Comparing this with the expression for $\mathbf{L}^2$ given by (5.10), we get

$$\frac{\mathbf{L}^2}{\hbar^2}\Theta(\theta)\Phi(\phi) = l(l+1)\Theta(\theta)\Phi(\phi).$$

Rewriting, we get

$$\mathbf{L}^2\,Y_{lm} = \hbar^2 l(l+1)\,Y_{lm}.$$

3. The radial part of the $2p$ wave function has the form (see Table (5.37)),

$$\psi_{21}(r) = Ar\exp\left(-\frac{r}{2a_0}\right).$$

The radial probability density is then,

$$p(r) = |\psi_{21}(r)|^2 r^2 = |A|^2 r^4 e^{-r/a_0}.$$

The peak is found by equating the r-derivative to zero,

$$\frac{dp(r)}{dr} = |A|^2 r^3 e^{-r/a_0}\left[4 - r/a_0\right] = 0.$$

This gives a maximum at $r = 4a_0$.

4.

$$\begin{aligned}\int_0^\infty r^2 e^{-2r/a_0}\,dr &= \left[-\frac{a_0}{2} r^2 e^{-2r/a_0}\right]_0^\infty + a_0 \int_0^\infty r e^{-2r/a_0}\,dr\\ &= a_0 \int_0^\infty r e^{-2r/a_0}\,dr.\end{aligned}$$

Integrating again by parts we get,

$$\begin{aligned}a_0 \int_0^\infty r e^{-2r/a_0}\,dr &= \left[-\frac{a_0^2}{2} r e^{-2r/a_0}\right]_0^\infty + \frac{a_0^2}{2}\int_0^\infty e^{-2r/a_0}\,dr\\ &= \frac{a_0^2}{2}\int_0^\infty e^{-2r/a_0}\,dr\\ &= \frac{a_0^2}{2}\left[-\frac{a_0}{2} e^{-2r/a_0}\right]_0^\infty = \frac{a_0^3}{4}.\end{aligned}$$

5. See Fig. 5.2.

6. (a) n: energy, $n = 1, 2, 3, ...$
l: angular momentum, $l = 0, 1, 2, ...(n-1)$
m: z-component of angular momentum, $m = -l, -l+1, ..., l$
The other quantum number that is needed is spin.

(b)

$$A^2 \int_0^\infty dr \int_0^{2\pi} d\phi \int_0^\pi d\theta |e^{-r/a_0}|^2 r^2 \sin\theta = 1.$$

Integrating over θ and ϕ gives,

$$4\pi A^2 \int_0^\infty r^2 e^{-2r/a_0}\,dr = 1.$$

Using the integral result supplied, the integral over r is $a_0^3/4$. Hence

$$4\pi A^2 \left(\frac{a_0^3}{4}\right) = 1$$

$$\text{i.e.} \qquad A = \frac{1}{\sqrt{\pi a_0^3}}.$$

(c)

$$\begin{aligned}\left\langle \frac{1}{r^2} \right\rangle &= \int_0^\infty dr \int_0^{2\pi} d\phi \int_0^\pi d\theta\, \psi_{100}^* \frac{1}{r^2} \psi_{100}\, r^2 \sin\theta \\ &= \frac{1}{\pi a_0^3} \int_0^{2\pi} d\phi \int_0^\pi d\theta \sin\theta \int_0^\infty e^{-2r/a_0}\, dr \\ &= \frac{4}{a_0^3} \left[\frac{-a_0}{2} e^{-2r/a_0} \right]_0^\infty = \frac{2}{a_0^2}.\end{aligned}$$

(d) This is because there is an additional geometric factor of r^2 in the radial probability density,

$$P(r) = r^2 |\psi_{nl}(r)|^2.$$

So, although the wave function is a maximum at $r = 0$, the radial density is zero. The two effect compete to give a maximum at some non-zero value of r.

7.

$$\text{Prob} = \frac{\int_0^b r^2 |\psi_{nl}(r)|^2\, dr}{\int_0^\infty r^2 |\psi_{nl}(r)|^2\, dr},$$

where $b = 10^{-15}$m.

$$\Longrightarrow \text{Prob} = \frac{\int_0^b r^2\, e^{-2r/a_0}\, dr}{\int_0^\infty r^2\, e^{-2r/a_0}\, dr},$$

Now, since $2b/a_0 \approx 4 \times 10^{-5} \ll 1$, we can approximate $e^{-r/a_0} \approx 1$ in the top line. This gives

$$\begin{aligned}\Longrightarrow \text{Prob} &= \frac{\int_0^b r^2\, dr}{\int_0^\infty r^2\, e^{-2r/a_0}\, dr}, \\ &= \frac{b^3/3}{\left(\frac{a_0}{2}\right)^3 2!} = \frac{4}{3} \left(\frac{b}{a_0} \right)^3 \approx 10^{-14}.\end{aligned}$$

12.5. Chapter 6

1. (a) The energy levels for non-interacting particles in a one-dimensional infinite square well are,

$$E_n = \frac{\pi^2 \hbar^2}{2ML^2} n^2,$$

where $n = 1, 2, 3, \ldots$ Two electrons (with opposite spin) can occupy each level, so the ground state consists of the 11th electron populating the $n = 6$ level. It, therefore, has energy

$$E_6 = \frac{\pi^2(1.055 \times 10^{-34})^2}{2(9.11 \times 10^{-31})(10^{-9})^2} 36 \approx 2.2 \times 10^{-18}\,\mathrm{J} \approx 13.6\,\mathrm{eV}.$$

(b) The lowest energy level that an electron can be excited to is $n = 6$ since all the others are full. So the energy required to excite an electron from the $n = 1$ level is

$$E = E_6 - E_1 = \frac{\pi^2(1.055 \times 10^{-34})^2}{2(9.11 \times 10^{-31})(10^{-9})^2}(6^2 - 1^2) \approx 13.2\,\mathrm{eV}.$$

2. Singlet:

$$\psi(x_1, x_2) = [\psi_p(x_1)\psi_q(x_2) + \psi_p(x_2)\psi_q(x_1)]\,(\chi_1(\uparrow)\chi_2(\downarrow) - \chi_1(\downarrow)\chi_2(\uparrow))\,.$$

Triplet:

$$\psi(x_1, x_2) = [\psi_p(x_1)\psi_q(x_2) - \psi_p(x_2)\psi_q(x_1)] \times \begin{cases} (\chi_1(\uparrow)\chi_2(\uparrow)) \\ (\chi_1(\uparrow)\chi_2(\downarrow) + \chi_1(\downarrow)\chi_2(\uparrow)) \\ (\chi_1(\downarrow)\chi_2(\downarrow)) \end{cases}$$

3. The average energy can be written as

$$\begin{aligned}\bar{E} &= \frac{\int_0^\infty E\,g(E)\,n(E)\,dE}{\int_0^\infty g(E)\,n(E)\,dE} \\ &= \left(\int_0^\infty \frac{E^{3/2}}{e^{(E-E_F)/kT}+1}dE\right) \div \left(\int_0^\infty \frac{E^{1/2}}{e^{(E-E_F)/kT}+1}dE\right) \\ &\approx \frac{\int_0^{E_F} E^{3/2}\,dE}{\int_0^{E_F} E^{1/2}\,dE} \\ &= \frac{\frac{2}{5}E_F^{5/2}}{\frac{2}{3}E_F^{3/2}} = \frac{3}{5}E_F.\end{aligned}$$

4. (a) The density of the conduction electrons can be calculated as follow:

$$\begin{aligned}\frac{N}{V} &= \frac{8.94\,\mathrm{g/cm^3}}{63.54\,\mathrm{g/mol}} \times (6.02 \times 10^{23}\mathrm{electrons/mol}) \\ &= 8.47 \times 10^{22}\,\mathrm{cm^{-3}} \\ &= 8.47 \times 10^{28}\,\mathrm{m^{-3}}.\end{aligned}$$

(b) From Eq. (6.37), the Fermi energy is given by

$$E_F = \frac{\hbar^2}{2M_e}\left(\frac{3\pi^2 N}{V}\right).$$

Substituting in values, including the value for N/V obtained in part (a), we get

$$E_F \approx 1.13 \times 10^{-18}\,\mathrm{J} = 7.05\,\mathrm{eV}.$$

(c) From Eq. (6.33), the Fermi velocity is given by

$$v_F = \sqrt{2E_F/M_e}.$$

Substituting in values we obtain

$$v_F \approx 1.58 \times 10^6\,\mathrm{m/s}.$$

12.6. Chapter 7

1. This can be found by adding up all the states that have $n = 4$, i.e. l can have values of $0, 1, 2, 3$ and these respectively have $1, 3, 5, 7$ values of m, i.e. 16 levels. However, since electrons are spin-1/2 particles, each level can be occupied by two opposite spin states. So the total number is $2 \times 16 = 32$.

2. For a given principal quantum number, n, we know that

$$\begin{aligned} l &= 0, 1, \cdots, (n-1) \\ m &= -l, -l+1, \cdots, l. \end{aligned}$$

   Also, there are two spin states, $m_s = \pm 1/2$, since an electron has spin-half. Adding these up, we get

$$\begin{aligned} 2 \times \sum_{l=0}^{n-1} 1 \sum_{m=-l}^{l} 1 &= 2 \times \sum_{l=0}^{n-1} (2l+1) \\ &= 4 \sum_{l=0}^{n-1} l + 2 \sum_{l=0}^{n-1} 1 \\ &= 4\left(\frac{n(n-1)}{2}\right) + 2n = 2n^2. \end{aligned}$$

3. The number of electrons gives the atomic number of an atom. In this case there are 20 electrons and so the atomic number is 20, i.e. calcium.

4. The energy levels in hydrogen are given by

$$E_n = -\frac{13.6\ \mathrm{eV}}{n^2} \qquad\qquad n = 1, 2, 3, \ldots$$

The longest wavelength in the absorption spectrum of ground state hydrogen corresponds to a transition to the next highest level, i.e. $n = 1 \rightarrow n = 2$. So the energy difference is $E_2 - E_1 = -13.6\text{eV}(1/4 - 1) = 10.2\,\text{eV}$. The wavelength of the transition is

$$\lambda = \frac{hc}{E} = \frac{(6.63 \times 10^{-34})(3 \times 10^8)}{10.2 \times 1.6 \times 10^{-19}} \approx 122\,\text{nm}.$$

5. (a) Possible, excited
   (b) Not possible
   (c) Possible, ground state
   (d) Possible, ground state
   (e) Possible, excited

6. The minimum potential energy of a pair of ions is

$$U(r = r_0) = -\frac{\alpha e^2}{4\pi\epsilon_0 r_0} + \frac{\beta}{r_0^\gamma},$$

where $r = r_0$ is the equilibrium separation. We can find an expression for r_0 by solving

$$\left.\frac{dU}{dr}\right|_{r=r_0} = 0.$$

This gives

$$\frac{\alpha e^2}{4\pi\epsilon_0 r_0^2} - \frac{\gamma\beta}{r_0^{\gamma+1}} = 0$$

$$\text{i.e.} \qquad \frac{\beta}{r_0^\gamma} = \frac{1}{\gamma}\frac{\alpha e^2}{4\pi\epsilon_0 r_0}.$$

Substituting this into the expression for $U_{\text{total}}(r = r_0)$, we get

$$U_{\min} = -\frac{\alpha e^2}{4\pi\epsilon_0 r_0}\left(1 - \frac{1}{\gamma}\right).$$

Of course, we could instead eliminate r_0 from this expression and write the minimum potential energy in terms of α, β and γ. In this case, we get

$$U_{\min} = -\left(\frac{\alpha e^2}{4\pi\epsilon_0}\right)^{\frac{\gamma}{\gamma-1}}\left(\frac{1}{\gamma\beta}\right)^{\frac{1}{\gamma-1}}\left(1 - \frac{1}{\gamma}\right).$$

7. It takes about 5.14 eV to ionise Na, and Cl releases 3.62 eV when it receives an extra electron. Overall that means that it takes:

$$E = (5.14 - 3.62)\text{eV} = 1.52\text{eV}$$

to create Na^+ and Cl^- form sodium and chlorine atoms. This process of creating an ionic bond becomes possible when it it energetically favourable, i.e. when the potential energy gained by the attraction of the two oppositely charged ions outweighs the energy cost in creating the ions. Since the electrostatic potential energy decreases as the ions get closer together, this enables us to find a maximum bond length purely from energy considerations.

Setting the magnitude of the potential energy equal to the cost of creating the ions, E, we get

$$E = \frac{e^2}{4\pi\epsilon_0 r}$$
$$\Longrightarrow r = \frac{e^2}{4\pi\epsilon_0 E}.$$

Substituting in values, we get:

$$\begin{aligned} r &= \frac{(1.6 \times 10^{-19}\,\mathrm{C})^2 \times (8.99 \times 10^9\,\mathrm{Nm^2C^{-2}})}{1.52 \times 1.6 \times 10^{-19}\,\mathrm{J}} \\ &\approx 9.5 \times 10^{-10}\,\mathrm{m} \\ &= 0.95\,\mathrm{nm}. \end{aligned}$$

This is the upper bound for the bond length of NaCl. For comparison, the observed bond length is about $r = 0.236$ nm, which is consistent with our result.

8. We know that the potential energy is

$$U(r) = -\frac{\alpha e^2}{4\pi\epsilon_0 r} + \frac{\beta}{r^\gamma},$$

and from exercise 6 above we found

$$\beta = \frac{1}{\gamma}\frac{\alpha e^2}{4\pi\epsilon_0} r_0^{\gamma-1}.$$

Substituting, we get

$$U(r) = -\frac{\alpha e^2}{4\pi\epsilon_0 r}\left[1 - \frac{1}{\gamma}\left(\frac{r_0}{r}\right)^{\gamma-1}\right].$$

Now, the force on the ion is given by

$$\begin{aligned} F(r) &= -\frac{dU(r)}{dr} \\ &= -\frac{\alpha e^2}{4\pi\epsilon_0 r^2}\left[1 - \left(\frac{r_0}{r}\right)^{\gamma-1}\right]. \end{aligned}$$

At a displacement $r = r_0 + x$, the force can be written as

$$F(r_0 + x) = -\frac{\alpha e^2}{4\pi\epsilon_0 r_0^2}\left[\left(1+\frac{x}{r_0}\right)^{-2} - \left(1+\frac{x}{r_0}\right)^{-(\gamma+1)}\right].$$

Using the fact that x is small, i.e. $x/r_0 \ll 1$, we can approximate the force by retaining only the first two terms in the binomial expansions of the two terms in the square brackets. This gives

$$\begin{aligned} F(r_0 + x) &= -\frac{\alpha e^2}{4\pi\epsilon_0 r_0^2}\left[1 - 2\frac{x}{r_0} - \left(1 - (\gamma+1)\frac{x}{r_0}\right)\right] \\ &= -\frac{\alpha e^2}{4\pi\epsilon_0 r_0^3}\left[(\gamma - 1)x\right]. \end{aligned}$$

This is in the form of Hooke's law

$$F = -Kx,$$

where the constant K is given by

$$K = \frac{\alpha e^2}{4\pi\epsilon_0 r_0^3}(\gamma - 1).$$

12.7. Chapter 8

1. The first thing to note is that the basis states are orthogonal since,

$$\langle\phi_2|\phi_1\rangle = \beta\alpha - \alpha\beta = 0,$$

and normalised because $|\alpha|^2 + |\beta|^2 = 1$. This means we can write the identity operator as

$$I = |\phi_1\rangle\langle\phi_1| + |\phi_2\rangle\langle\phi_2|.$$

Operating on state $|\psi\rangle$ with this gives

$$\begin{aligned} I|\psi\rangle &= (|\phi_1\rangle\langle\phi_1| + |\phi_2\rangle\langle\phi_2|)(a|0\rangle + b|1\rangle) \\ &= (a\alpha^* + b\beta^*)|\phi_1\rangle + (a\beta - b\alpha)|\phi_2\rangle. \end{aligned}$$

2. This can be confirmed by direct substitution

$$\begin{aligned} I &= |+\rangle\langle+| + |-\rangle\langle-| \\ &= \frac{1}{2}(|\uparrow\rangle + |\downarrow\rangle)(\langle\uparrow| + \langle\downarrow|) + \frac{1}{2}(|\uparrow\rangle - |\downarrow\rangle)(\langle\uparrow| - \langle\downarrow|) \\ &= \frac{1}{2}(2|\uparrow\rangle\langle\uparrow| + 2|\downarrow\rangle\langle\downarrow| + |\uparrow\rangle\langle\downarrow| + |\downarrow\rangle\langle\uparrow| - |\uparrow\rangle\langle\downarrow| - |\downarrow\rangle\langle\uparrow|) \\ &= |\uparrow\rangle\langle\uparrow| + |\downarrow\rangle\langle\downarrow|. \end{aligned}$$

3. Substituting the state $|\psi(t)\rangle$ into both sides of the time-dependent Schrödinger equation we get

$$i\hbar\frac{\partial}{\partial t}\sum_{i=0}^{n-1}\alpha_i\, e^{-iE_it/\hbar}\,|\psi_i\rangle = \hat{H}\sum_{i=0}^{n-1}\alpha_i\, e^{-iE_it/\hbar}\,|\psi_i\rangle$$
$$i\hbar\sum_{i=0}^{n-1}\alpha_i\left(\frac{\partial}{\partial t}e^{-iE_it/\hbar}\right)|\psi_i\rangle = \sum_{i=0}^{n-1}\alpha_i\, e^{-iE_it/\hbar}\,(\hat{H}|\psi_i\rangle)$$
$$i\hbar\sum_{i=0}^{n-1}\alpha_i\left(\frac{-iE_i}{\hbar}\right)e^{-iE_it/\hbar}\,|\psi_i\rangle = \sum_{i=0}^{n-1}\alpha_i\, e^{-iE_it/\hbar}\,E_i|\psi_i\rangle$$
$$\sum_{i=0}^{n-1}\alpha_i\, e^{-iE_it/\hbar}\,E_i|\psi_i\rangle = \sum_{i=0}^{n-1}\alpha_i\, e^{-iE_it/\hbar}\,E_i|\psi_i\rangle.$$

We see that the left and right hand sides are equal and so $|\psi(t)\rangle$ is indeed a solution of the time-dependent Schrödinger equation.

4. (a) $|R\rangle$ and $|L\rangle$ are orthogonal since

$$\begin{aligned}\langle R|L\rangle &= \frac{1}{2}((\langle H| - i\langle V|)(|H\rangle - i|V\rangle))\\ &= \frac{1}{2}(\langle H|H\rangle - i\langle H|V\rangle - i\langle V|H\rangle - \langle V|V\rangle)\\ &= \frac{1}{2}(1-1) = 0.\end{aligned}$$

(b) $|R\rangle$ and $|L\rangle$ are normalised since

$$\begin{aligned}\langle R|R\rangle &= \frac{1}{2}((\langle H| - i\langle V|)(|H\rangle + i|V\rangle))\\ &= \frac{1}{2}(\langle H|H\rangle + i\langle H|V\rangle - i\langle V|H\rangle + \langle V|V\rangle)\\ &= \frac{1}{2}(1+1) = 1\\ \langle L|L\rangle &= \frac{1}{2}((\langle H| + i\langle V|)(|H\rangle - i|V\rangle))\\ &= \frac{1}{2}(\langle H|H\rangle - i\langle H|V\rangle + i\langle V|H\rangle + \langle V|V\rangle)\\ &= \frac{1}{2}(1+1) = 1.\end{aligned}$$

(c) One way to find γ and δ is as follows:

$$\begin{aligned}\gamma = \langle R|\psi\rangle &= \alpha\langle R|H\rangle + \beta\langle R|V\rangle\\ &= \frac{1}{\sqrt{2}}(\alpha - i\beta)\\ \delta = \langle L|\psi\rangle &= \alpha\langle L|H\rangle + \beta\langle L|V\rangle\\ &= \frac{1}{\sqrt{2}}(\alpha + i\beta).\end{aligned}$$

5. (a) Possible outcomes of an energy measurement: E_1, E_2, or E_3.

   (b) Expectation value:

$$\begin{aligned}\langle\psi|H|\psi\rangle &= (\alpha^*\langle 1| + \beta^*\langle 2| + \gamma^*\langle 3|)(E_1|1\rangle\langle 1| + E_2|2\rangle\langle 2| + E_3|3\rangle\langle 3|) \\ &\qquad\times(\alpha|1\rangle + \beta|2\rangle + \gamma|3\rangle) \\ &= (\alpha^*\langle 1| + \beta^*\langle 2| + \gamma^*\langle 3|)(E_1\alpha|1\rangle + E_2\beta|2\rangle + E_3\gamma|3\rangle) \\ &= |\alpha|^2E_1 + |\beta|^2E_2 + |\gamma|^2E_3.\end{aligned}$$

   (c) Using the same technique as in (b), we get

$$\langle\psi|H|\psi\rangle = |\alpha|^2E_1 + |\beta|^2E_2 + |\gamma|^2E_3 + \Omega(\alpha^*\beta + \beta^*\alpha).$$

6. (a) Eigenvalues:

$$\begin{aligned}A|\psi_i\rangle &= (|1\rangle\langle 2| + |2\rangle\langle 1|)(c_1|1\rangle + c_2|2\rangle) \\ &= c_2|1\rangle + c_1|2\rangle\end{aligned}$$

   Now, we require the right hand side to be equal to,

$$A|\psi_i\rangle = \alpha_i|\psi_i\rangle = \alpha_i(c_1|1\rangle + c_2|2\rangle).$$

   This means we have the conditions:

$$\begin{aligned}\alpha_i c_1 &= c_2 \\ \alpha_i c_2 &= c_1.\end{aligned}$$

   These are easily solved by substitution to give $\alpha_i^2 = 1$, i.e. the two eigenvalues are $\alpha_i = \pm 1$.

   (b) The eigenvectors are given by

$$|\psi_i\rangle = \begin{pmatrix} c_1 \\ c_2 \end{pmatrix} = \begin{pmatrix} c_1 \\ \alpha_i c_1 \end{pmatrix},$$

   where we have used the conditions found in (a). Normalisation gives

$$\begin{aligned}1 = \langle\psi|\psi\rangle &= |c_1|^2 + |\alpha_i|^2|c_1|^2 \\ &= 2|c_1|^2,\end{aligned}$$

   since we know that $|\alpha_i|^2 = 1$. This means $c_1 = 1/\sqrt{2}$ and the eigenvectors corresponding to the two eigenstates are:

$$\begin{aligned}\alpha_i = +1 \quad &\Longrightarrow \quad |\psi_i\rangle = \frac{1}{\sqrt{2}}\begin{pmatrix} 1 \\ 1 \end{pmatrix} \\ \alpha_i = -1 \quad &\Longrightarrow \quad |\psi_i\rangle = \frac{1}{\sqrt{2}}\begin{pmatrix} 1 \\ -1 \end{pmatrix}.\end{aligned}$$

7. The annihilation operator can be written as

$$\begin{aligned} a &= \sqrt{\frac{M\omega}{2\hbar}}\left(\hat{x} + \frac{i}{M\omega}\hat{p}\right) \\ &= \frac{1}{\sqrt{2}b}\left(\hat{x} + b^2\frac{d}{dx}\right), \end{aligned}$$

where $b = \sqrt{\hbar/(M\omega)}$. So $a\psi_0(x) = 0$ gives

$$\left(\hat{x} + b^2\frac{d}{dx}\right)\psi_0(x) = 0,$$

$$\text{i.e.} \qquad x\psi_0(x) + b^2\frac{d\psi_0(x)}{dx} = 0.$$

The solution to this equation is

$$\psi_0(x) = C\exp\left(\frac{-x^2}{2b^2}\right),$$

which is the ground state wave function. The constant, C can be found by normalisation.

8. The first excited state is

$$\psi_1(x) = \left(\frac{2}{\sqrt{\pi}b^3}\right)^{1/2} xe^{-x^2/2b^2},$$

where $b = \sqrt{\hbar/(M\omega)}$. So $a^\dagger\psi_1(x)$ gives,

$$\begin{aligned} a^\dagger\psi_1(x) &= \frac{1}{\sqrt{2}b}\left(\hat{x} - b^2\frac{d}{dx}\right)\left(\frac{2}{\sqrt{\pi}b^3}\right)^{1/2} xe^{-x^2/2b^2} \\ &= \left(\frac{1}{\sqrt{\pi}b}\right)^{1/2}\left[\frac{x^2}{b^2}e^{-x^2/2b^2} - e^{-x^2/2b^2} + \frac{x^2}{b^2}e^{-x^2/2b^2}\right] \\ &= \sqrt{2}\left(\frac{1}{2\sqrt{\pi}b}\right)^{1/2}\left(\frac{2x^2}{b^2} - 1\right)e^{-x^2/2b^2} \\ &= \sqrt{2}\,\psi_2(x), \end{aligned}$$

where the last step follows from Eq. (4.16).

9. We can use first-order perturbation theory since we know the ground state solution of $V(x)$, i.e.

$$\psi^{(0)}(x) = \left(\frac{1}{\sqrt{\pi}b}\right)^{1/2} e^{-x^2/2b^2},$$

where $b = \sqrt{\hbar/(M\omega)}$. The first-order correction to the ground state energy

$$E^{(1)} = \langle\psi^{(0)}|x^3|\psi^{(0)}\rangle,$$

is, therefore,

$$E^{(1)} = \frac{1}{\sqrt{\pi}b}\int_{-\infty}^{\infty} e^{-x^2/2b^2} x^3 e^{-x^2/2b^2}\, dx$$
$$= \frac{1}{\sqrt{\pi}b}\int_{-\infty}^{\infty} x^3 e^{-x^2/b^2}\, dx.$$

We can see that this integral vanishes by observing that the integrand is odd. This means that the first order correction to the ground state energy is zero. Note that this does not mean that the perturbation does not change the ground state energy merely that the *first order* correction is zero. If we were to extend the perturbation calculation to higher orders we would see a shift in the energy due to the perturbation.

10. We can use first-order perturbation theory since we know the ground state solution of $V(x)$, i.e.

$$\psi^{(0)}(x) = \sqrt{\frac{2}{L}} \sin\left(\frac{\pi x}{L}\right).$$

The first-order correction to the ground state energy

$$E^{(1)} = \langle \psi^{(0)} | e^{\hat{x}} | \psi^{(0)} \rangle,$$

is, therefore,

$$E^{(1)} = \frac{2}{L}\int_0^L \sin\left(\frac{\pi x}{L}\right) e^x \sin\left(\frac{\pi x}{L}\right) dx$$
$$= \frac{2}{L}\int_0^L e^x \sin^2\left(\frac{\pi x}{L}\right) dx.$$

Using the integral identity given, this can be evaluated as

$$E^{(1)} = \frac{4\pi^2(e^L - 1)}{L(4\pi^2 + L^2)},$$

and substituting $L = 2\pi$, this reduces to

$$E^{(1)} = \frac{e^{2\pi} - 1}{4\pi}.$$

This means that, to first order, the ground state energy of the perturbed potential is

$$E \approx \frac{\pi^2\hbar^2}{2M(2\pi)^2} + \lambda\frac{e^{2\pi} - 1}{4\pi}$$
$$= \frac{\hbar^2}{8M} + \lambda\frac{e^{2\pi} - 1}{4\pi}.$$

12.8. Chapter 9

1. Neither the captain nor his deputy are right. The explosion and acceleration are spacelike separated events. This means that they are outside each other light cones, which means that there is no unique order for them for all observers. The order can be any, depending on your motion.

2. Here the captain is wrong. The preservation of causality is the whole point of Einstein's theory. Cause must precede effect in all frames. Causing someone to die in your inertial frame, just causes them to die in all other frames of reference. In order to see the two events reversed, one must travel faster than the speed of light, which is impossible according to relativity. The gamma coefficient becomes imaginary and this physically does not make sense.

3. Let us start by finding an expression for v_n. If we make the substitution $\tilde{r_n} = \gamma r_n$, Eqs. (9.42) and (9.43) can be written as

$$\begin{aligned} n\hbar &= M_e v_n \tilde{r_n} \\ \frac{Ze^2}{4\pi\epsilon_0} &= M_e v_n^2 \tilde{r_n}. \end{aligned}$$

We can eliminate $\tilde{r_n}$ by, for example, dividing the lower equation by the upper one. This gives an expression for v_n

$$v_n = \frac{Ze^2}{4\pi\epsilon_0 \hbar n}.$$

Comparing this with Eq. (2.48), we see that, interestingly, the relativistic expression for the velocity of the electrons in the Bohr orbits is identical to the non-relativistic result.

Using this result, the γ factor is given by

$$\begin{aligned} \frac{1}{\gamma} &= \sqrt{1 - \left(\frac{Ze^2}{4\pi\epsilon_0 \hbar c}\right)^2 \frac{1}{n^2}} \\ &= \sqrt{1 - \frac{\alpha^2}{n^2}}. \end{aligned}$$

Then from Eq. (9.42), we have

$$\begin{aligned} r_n &= \frac{n\hbar}{\gamma M_e v_n} = \frac{n\hbar}{M_e}\left(\frac{4\pi\epsilon_0 \hbar n}{Ze^2}\right)\sqrt{1 - \frac{\alpha^2}{n^2}} \\ &= \frac{4\pi\epsilon_0 \hbar^2}{M_e e^2} n\sqrt{n^2 - \alpha^2} \\ &= a_0 n\sqrt{n^2 - \alpha^2}, \end{aligned}$$

where a_0 is the Bohr radius.

4. Let us first calculate Bohr's velocity of the electron in the lowest orbit. It is given by

$$v_1 = \frac{e^2}{4\pi\epsilon_0\hbar} \approx 3 \times 10^6 \text{m/s}.$$

This is one percent of the speed of light, which allows us to calculate

$$\gamma = \frac{1}{\sqrt{1 - v^2/c^2}} \approx 1 + \frac{1}{2}10^{-4}.$$

The mass is changed through

$$M = \gamma M_0,$$

which has a knock on effect on the change in the wavelength through the inverse Rydberg constant. The difference between the relativistic and non-relativistic $\Delta\lambda$ is, therefore, only one half in ten thousand – very small indeed.

5. The fraction of muons remaining after the travel time t is

$$\left(\frac{1}{2}\right)^{t/t'_{1/2}} = \frac{700}{1000} = 0.7,$$

where t is the travel time in the frame of the detectors, i.e. $t = d/v$, where d is the distance travelled by the muons and v is the speed of the muons. The half-life of the muons in the same frame, i.e. that of the detectors, is $t'_{1/2}$ and this is related to the half-life of the muons in their rest frame, $t_{1/2}$ by

$$t'_{1/2} = \gamma t_{1/2}.$$

Taking the log of both side of the first equation, we get

$$t = t'_{1/2}\frac{\log(0.7)}{\log(0.5)}.$$

Making the substitutions discussed above gives

$$\frac{v}{d} = \frac{t_{1/2}}{\sqrt{1 - v^2/c^2}}\frac{\log(0.7)}{\log(0.5)}.$$

This can be solved for v to give

$$v = c\left[1 + \frac{c^2t_{1/2}^2}{d^2}\left(\frac{\log(0.7)}{\log(0.5)}\right)^2\right]^{-1/2}.$$

Substituting in numerical values: $c = 3 \times 10^8$m/s, $d = 2000$m, $t_{1/2} = 1.56\mu$s, we get

$$v \approx 0.993c.$$

6. The energy released in the reaction

$$^{235}\mathrm{U} + \mathrm{n} \longrightarrow {}^{141}\mathrm{Ba} + {}^{92}\mathrm{Kr} + 3\mathrm{n}$$

is found by calculating how much smaller the mass of the fragments on the right hand side is relative to the left hand side and then using $E = Mc^2$. Using the given atomic masses: ^{235}U= 235.043925u; ^{141}Ba=140.914406u; ^{92}Kr=91.926156u; n=1.008665u, the change in mass is:

$$\begin{aligned}\Delta M &= (235.043925)u - (140.914406 + 91.926156 + 2 \times 1.008665)u \\ &= 0.186033u.\end{aligned}$$

The atomic mass is given in convenient units by $u = 931.5\ \mathrm{MeV}/c^2$. So the energy released is

$$\begin{aligned}E &= 0.186033 \times 931.5\,\mathrm{MeV}/c^2 \times c^2 \\ &\approx 173\,\mathrm{MeV}.\end{aligned}$$

12.9. Chapter 10

1. The electromagnetic field is classically a wave described by six numbers at every point in space and time. These numbers can be specified simultaneously and the values of both the electric and magnetic field can in principle be determined exactly and simultaneously. When the field is quantised this is no longer possible. In fact, the electric and magnetic fields become operators which no longer commute, i.e. they are no longer simultaneously measurable. Writing the total classical energy in the field we get

$$W = \frac{1}{2}\int (\epsilon E^2 + \mu H^2) dV,$$

but $E \propto x \cos \omega t/V$ and $B \propto p\omega \sin \omega t/V$, so that $W = (p^2 + \omega^2 x^2)/2$, which is a simple harmonic oscillator with unit mass.

2. The probability of obtaining n photons is $|\langle n|\alpha\rangle|^2$ and is

$$p_n = e^{-|\alpha|^2} \frac{|\alpha|^{2n}}{n!}$$

The average energy is

$$\langle E\rangle = \hbar\omega\langle\alpha|\left(\hat{a}^\dagger\hat{a} + \frac{1}{2}\right)|\alpha\rangle$$

$$= \hbar\omega\left(e^{-|\alpha|^2}\sum_n \frac{|\alpha|^{2n}}{n!}n + \frac{1}{2}\right)$$
$$= \hbar\omega\left(e^{-|\alpha|^2}|\alpha|^2\sum_n \frac{|\alpha|^{2(n-1)}}{(n-1)!} + \frac{1}{2}\right)$$
$$= \hbar\omega\left(|\alpha|^2 + \frac{1}{2}\right)$$

3. The physical meaning of $|\alpha|^2$ is the average number of photons.

4. We now solve the Schrödinger equation to obtain the free evolution of this state

$$i\hbar\frac{\partial|n\rangle}{\partial t} = \hbar\omega\left(\hat{a}^\dagger\hat{a} + \frac{1}{2}\right)|n\rangle.$$

The solution is

$$|n(t)\rangle = e^{i\omega(n+1/2)t}|n\rangle,$$

Therefore,

$$|\alpha(t)\rangle = e^{-|\alpha|^2/2}\sum_n \frac{\alpha^n}{\sqrt{n!}}e^{i\omega(n+1/2)t}|n\rangle$$
$$= e^{i\omega t/2}e^{-|\alpha|^2/2}\sum_n \frac{(\alpha e^{i\omega t})^n}{\sqrt{n!}}|n\rangle$$
$$= e^{i\omega t/2}|\alpha e^{i\omega t}\rangle.$$

Therefore, the amplitude oscillates at frequency ω (the overall phase is not important in quantum mechanics).

The probability does not depend on this overall phase (whose mod square is equal to one). Therefore, it does not change with time.

5. The relativistic form of the kinetic energy is

$$T = \sqrt{p^2c^2 + M^2c^4} - Mc^2,$$

where the first term is the total relativistic energy and the second term is the rest energy of the electron. Expanding the square-root, we get

$$T = Mc^2\left[\sqrt{1 + \frac{p^2}{M^2c^2}} - 1\right]$$
$$= Mc^2\left[-1 + 1 + \frac{1}{2}\left(\frac{p^2}{M^2c^2}\right) - \frac{1}{8}\left(\frac{p^2}{M^2c^2}\right)^2 + \cdots\right]$$
$$\approx \frac{p^2}{2M} - \frac{p^4}{8M^3c^2}.$$

So the first-order correction to the kinetic energy due to relativistic effects is

$$H' = -\frac{p^4}{8M^3c^2}.$$

The first-order correction to the ground state energy of the hydrogen atom is, therefore,

$$E_0^{(1)} = -\frac{1}{8M^3c^2}\langle\psi_0^{(0)}|p^4|\psi_0^{(0)}\rangle,$$

where $|\psi_0^{(0)}\rangle$ is the unperturbed ground state described by the normalised wave function

$$\frac{1}{\sqrt{\pi}a_0^{3/2}}e^{-r/a_0}.$$

Now, we can make use of the Schrödinger equation,

$$\left(\frac{p^2}{2M} + V(r)\right)|\psi_0^{(0)}\rangle = E_0|\psi_0^{(0)}\rangle,$$

to write

$$p^2|\psi_0^{(0)}\rangle = 2M(E_0 - V(r))|\psi_0^{(0)}\rangle.$$

This allows us to rewrite the relativistic correction as

$$\begin{aligned} E_0^{(1)} &= -\frac{1}{8M^3c^2}\langle\psi_0^{(0)}|(2M)^2(E_0 - V(r))^2|\psi_0^{(0)}\rangle \\ &= -\frac{1}{2Mc^2}\left(E_0^2 - 2E_0\langle V\rangle + \langle V^2\rangle\right). \end{aligned}$$

We can now evaluate $\langle V\rangle$ and $\langle V^2\rangle$ directly:

$$\begin{aligned} \langle V\rangle &= 4\pi\int_0^\infty r^2\left(\frac{-e^2}{4\pi\epsilon_0 r}\right)\frac{1}{\pi a_0^3}e^{-2r/a_0}\,dr \\ &= \frac{-e^2}{4\pi\epsilon_0 a_0} \end{aligned}$$

(using the integral identity provided) and,

$$\begin{aligned} \langle V^2\rangle &= 4\pi\int_0^\infty r^2\left(\frac{-e^2}{4\pi\epsilon_0 r}\right)^2\frac{1}{\pi a_0^3}e^{-2r/a_0}\,dr \\ &= 2\left(\frac{e^2}{4\pi\epsilon_0 a_0}\right)^2. \end{aligned}$$

Substituting these results back and using

$$E_0 = \frac{-e^2}{8\pi\epsilon_0 a_0},$$

which is obtained form Eqs. 5.36 and 5.38, we obtain

$$\begin{aligned} E_0^{(1)} &= -\frac{1}{2Mc^2}\left(E_0^2 - 2E_0\langle V\rangle + \langle V^2\rangle\right) \\ &= -\frac{5}{8Mc^2}\left(\frac{e^2}{4\pi\epsilon_0 a_0}\right)^2, \end{aligned}$$

which is the first-order relativistic correction.

12.10. Chapter 11

1. The entropy is given by

$$S = -\mathrm{Tr}\{\rho \log_2 \rho\}.$$

This can be evaluated by finding the eigenvalues of the density matrix ρ, i.e. diagonalising the matrix. The eigenvalues are the solution to

$$\begin{aligned} &\det\begin{pmatrix} 1/2-\lambda & 1/4 \\ 1/4 & 1/2-\lambda \end{pmatrix} = 0 \\ &\Longrightarrow (1/2-\lambda)^2 - (1/4)^2 = 0. \end{aligned}$$

The solutions are $\lambda = 1/4, 3/4$ and so the entropy is

$$\begin{aligned} S &= -\frac{1}{4}\log_2\frac{1}{4} - \frac{3}{4}\log_2\frac{3}{4} \\ &= \log_2 4 - \frac{3}{4}\log_2 3 \approx 0.81. \end{aligned}$$

2. The density matrix for the state is

$$\rho = |a|^2|00\rangle\langle 00| + |b|^2|11\rangle\langle 11| + ab^*|00\rangle\langle 11| + a^*b|11\rangle\langle 00|.$$

If we trace out the second qubit, the reduced density matrix is

$$\rho_A = |a|^2|0\rangle\langle 0| + |b|^2|1\rangle\langle 1|.$$

This means that the entropy of entanglement is given by

$$E = -\mathrm{Tr}\{\rho_A \log_2 \rho_A\} = -|a|^2\log_2|a|^2 - |b|^2\log_2|b|^2.$$

This is maximised when $|a|^2 = |b|^2 = 1/2$. The entropy of entanglement is zero when either $|a| = 1$ and $|b| = 0$ or conversely $|a| = 0$ and $|b| = 1$. This is because the original state is then clearly a product state, i.e. it is separable.

3. (a) The density matrix is given by

$$\rho = \frac{1}{3}\left[|g\rangle\langle g| + |e\rangle\langle e| + \frac{1}{2}(|g\rangle + |e\rangle)(\langle g| + \langle e|)\right],$$

which can be written as

$$\rho = \frac{1}{3}\begin{pmatrix} 3/2 & 1/2 \\ 1/2 & 3/2 \end{pmatrix} = \begin{pmatrix} 1/2 & 1/6 \\ 1/6 & 1/2 \end{pmatrix}.$$

Diagonalising this, we find the eigenvalues are $\lambda = 1/3, 2/3$. The entropy of the state is, therefore,

$$\begin{aligned} S &= -\frac{1}{3}\log_2\frac{1}{3} - \frac{2}{3}\log_2\frac{2}{3} \\ &= \log_2 3 - 2/3 \approx 0.92. \end{aligned}$$

(b) After the measurement, the density matrix is

$$\rho_{\text{meas}} = \frac{1}{2}\left[|g\rangle\langle g| + |e\rangle\langle e|\right],$$

and so the entropy of this state is

$$S(\rho_{\text{meas}}) = \log_2 2 = 1.$$

(c) The entropy is larger after the measurement, i.e. $S(\rho_{\text{meas}}) > S$. This can be explained by the fact that noise is introduced by the measurement apparatus.

4. (a) The density matrix is

$$\rho = \frac{9}{10}|\Phi^+\rangle\langle\Phi^+| + \frac{1}{10}|00\rangle\langle 00|.$$

(b) Suppose that we wish to teleport the state $a|0\rangle + b|1\rangle$. We know that 90% of the time Alice and Bob share the maximally entangled state $|\Phi^+\rangle$ and so the teleportation protocol works perfectly (i.e. with unit fidelity). The other 10% of the time, Alice and Bob share the separable state $|00\rangle$. Let us consider what happens when we try and teleport with that. The initial state can be written as

$$\begin{aligned} (a|0\rangle + b|1\rangle)|00\rangle &= a|000\rangle + b|100\rangle \\ &= \frac{a}{\sqrt{2}}(|\Phi^+\rangle + |\Phi^-\rangle)|0\rangle + \frac{b}{\sqrt{2}}(|\Psi^+\rangle - |\Psi^-\rangle)|0\rangle. \end{aligned}$$

After Alice measures one of the Bell states, informs Bob of the result, and he makes the appropriate state rotation, the state that Bob is left with is

$$|a|^2|0\rangle\langle 0| + |b|^2|1\rangle\langle 1|.$$

Therefore, combining this result with what happens 90% of the time, i.e. when Alice and Bob share the Bell state $|\Phi^+\rangle$ and so achieve perfect teleportation, the overall teleported state is

$$\begin{aligned}\rho &= \frac{9}{10}(a|0\rangle + b|1\rangle)(a^*\langle 0| + b^*\langle 1|) + \frac{1}{10}\left(|a|^2|0\rangle\langle 0| + |b|^2|1\rangle\langle 1|\right) \\ &= |a|^2|0\rangle\langle 0| + |b|^2|1\rangle\langle 1| + \frac{9}{10}(ab^*|0\rangle\langle 1| + a^*b|1\rangle\langle 0|).\end{aligned}$$

This can, equivalantly, be written as

$$\rho = \begin{pmatrix} |a|^2 & 9ab^*/10 \\ 9a^*b/10 & |b|^2 \end{pmatrix}.$$

Bibliography

[1] Guthrie W.K.C. (1971). *Socrates*, Cambridge.

[2] Elitzur A.C. Vaidman L. (1993). Quantum-mechanical interaction-free measurements. *Foundations of Physics* **23**: 987–997.

[3] Kwiat P. *et al.* (1995). Interaction-free measurement. *Physical Review Letters* **74**: 4763–4766.

[4] Vedral V. (2005). *Modern Foundations of Quantum Optics*, Imperial College Press, London.

[5] Alpher R. A., Bethe H., Gamow G. (1948). The Origin of Chemical Elements. *Physical Review* **73**: 803-804.

[6] Foot C.J. (2005). *Atomic Physics*, Oxford.

[7] Phillips W.D. (1998). Nobel Lecture: Laser cooling and trapping of neutral atoms. *Review of Modern Physics* **70**: 721–741.

[8] Bransden B.H., Joachain C.J. (1989). *Introduction to Quantum Mechanics*, Longman, Harlow.

[9] Bailey J. *et al.* (1977). Measurements of relativistic time dilatation for positive and negative muons in a circular orbit. *Nature* **268**: 301–305.

[10] Lifshitz E.M., Pitaevskii L.P. (1980). *Statistical Physics Part 2: Landau and Lifshitz Course of Theoretical Physics*, Reed, Oxford.

[11] Nielsen M., Chuang I. (2000). *Quantum Computation and Quantum Information*, Cambridge.

[12] Vedral V. (2006). *Introduction to Quantum Information Science*, Oxford.

[13] Plenio M., Vedral V. (1998). Teleportation, entanglement and thermodynamics in the quantum world. *Contemporary Physics* **39**: 431–446.

[14] Linden N., Massar S., Popescu S. (1998). Purifying Noisy Entanglement Requires Collective Measurements. *Physical Review Letters* **81**: 3279–3282.

[15] Vedral V. (2010). *Decoding Reality: The Universe as Quantum Information*, Oxford.

Index